Lewis Carroll's Cats and Rats... and Other Puzzles with Interesting Tails

LEWIS CARROLL'S CATS AND RATS... AND OTHER PUZZLES WITH INTERESTING TAILS

Yossi Elran

World Scientific

NEW JERSEY · LONDON · SINGAPORE · BEIJING · SHANGHAI · HONG KONG · TAIPEI · CHENNAI · TOKYO

Published by

World Scientific Publishing Co. Pte. Ltd.
5 Toh Tuck Link, Singapore 596224
USA office: 27 Warren Street, Suite 401-402, Hackensack, NJ 07601
UK office: 57 Shelton Street, Covent Garden, London WC2H 9HE

Library of Congress Cataloging-in-Publication Data
Names: Elran, Yossi, 1967–   author.
Title: Lewis Carroll's cats and rats ... and other puzzles with interesting tails / Yossi Elran.
Description: Hackensack, New Jersey : World Scientific, [2021] |
    Includes bibliographical references.
Identifiers: LCCN 2021009903 (print) | LCCN 2021009904 (ebook) |
    ISBN 9789811233968 (hardcover) | ISBN 9789811235641 (paperback) |
    ISBN 9789811233975 (ebook for institutions) | ISBN 9789811233982 (ebook for individuals)
Subjects: LCSH: Mathematical recreations. | Mathematical recreations--History. |
    Mathematics--Problems, exercises, etc.
Classification: LCC QA95 .E395 2021 (print) | LCC QA95 (ebook) | DDC 793.73--dc23
LC record available at https://lccn.loc.gov/2021009903
LC ebook record available at https://lccn.loc.gov/2021009904

British Library Cataloguing-in-Publication Data
A catalogue record for this book is available from the British Library.

For any available supplementary material, please visit
https://www.worldscientific.com/worldscibooks/10.1142/12198#t=suppl

Printed in Singapore

To Michal

Preface

When I was growing up, I had a great passion for two things: the Babylonian Talmud, an exceptional, collective, intellectual work of my ancient, rabbinical ancestors, and recreational math — the math behind games, puzzles and the arts. In fact, I combined the two by sitting at the back of the classroom in Talmud lessons, while reading Martin Gardner's recreational math books beneath the desk. Luckily for me, I never got caught, though I suspect my teachers "turned a blind eye". My passion never ceased, and though I have done many things since then, I have never stopped learning the Talmud, and dabbling with recreational math. Throughout the years, particularly since I joined the Weizmann Institute of Science and the Davidson Institute of Science Education over twenty years ago, I have taught recreational math on many platforms — math circles, afternoon clubs, online courses, to name just a few. In many of them, I have used a methodology based on the "associative learning" model of the Talmud. The idea is, rather like conversation. You start off with a given concise text, which you then analyze and tackle from different angles, often straying off into "other worlds", and finally returning with many insights at the end.

Lewis Carroll's Cats and Rats ... and Other Puzzles with Interesting Tails was written just like that. On the one hand, it is a classic recreational math book, along the lines of the works of Martin Gardner and others, touching on many new ideas, explanations and trends in recreational math. On the other hand, the book has a unique structure: each chapter starts with a single puzzle, problem or magic trick. It then explores the math behind the puzzle and its solution, and exposes many hidden math themes and connected issues, along with some of the stories, trivia, quirks and oddities of their history and the mathematicians behind them. At the end of each chapter, I've given some follow-up puzzles, for you to enjoy and solve, and (separately) their solutions. I hope you enjoy — and benefit from — reading this book and that you notice the associative way in which it was written. If you are a teacher, I would in particular suggest you try teaching your subjects associatively. I find that "steering off course" in many directions keeps the students "gripped" and they are able to stay attentive and, even more importantly, enjoy math!

This book would not have been possible without the help of many people. First and foremost, World Scientific Publishing, especially Rochelle Kronzek, without whose support, professionalism and encouragement I would never have finished this book; my editors, Ying-Oi Chiew and Lai-Fun Kwong who were always very accommodating and helpful with all my requests. I would like to thank Prof. Oded Margalit, Prof. Rami Band and Skona Brittain for their priceless advice. I would also like to thank Dr. Gyora Benedek for the Hidato puzzles in this book. The Davidson Institute of Science Education at the Weizmann Institute of Science has been my professional home and family for a long time. I am very grateful to them, especially my CEO, Dr. Liat Ben-David for their ongoing support and help.

Finally, I would like to thank my dear, loving wife Michal, herself a professional computer scientist, linguist, recreational mathematician and online learning professional, who has been, and always is, such an amazing inspiration for me, and is a constant source of support, comfort and encouragement to me and to our children, Gilad, Tamar, Racheli, Sharon and Hadas. I love you all so much! Without you, none of this could have ever, ever, happened.

Yossi Elran, 2021

Contents

Lewis Carroll's Cats and Rats

Introduction

One of my favorite puzzles is little known, but quite old. It is attributed to Lewis Carroll, the famous author of *Alice in Wonderland* and other wonderful books, but originated much earlier — with a surprising story behind it.

The Puzzle

If 6 cats kill 6 rats in 6 minutes, how many will be needed to kill 100 rats in 50 minutes? (No animals were hurt in preparing this puzzle ...)

Where to Start?

You might need a hint to solve some of the puzzles in this book, however this puzzle seems to be straightforward. All the information that is needed is in the question, in particular the "killing rate" of six rats.

Before you go any further, why don't you try and solve the puzzle right now? What was your answer? Was it 9, 10, 11, 12 or 13?

Solving the Puzzle

Surprisingly, 12, 13 and 14 are *all* correct answers! How is this possible? It all depends on the reasoning. Let's see for ourselves.

- The answer is **12**:

    If 6 cats kill 6 rats in 6 minutes, then 6 cats collectively kill one rat every minute. In 50 minutes, the 6 cats can kill 50 rats. So to kill 100 rats, twice the number of cats is needed — 12 cats.

  or:

    If 6 cats kill 6 rats in 6 minutes, then a group of 3 cats kills a rat in 2 minutes. In 50 minutes, every group of 3 cats can kill 25 rats because $50 \div 2 = 25$. To kill 100 rats in 50 minutes, 4 triplets of cats are needed because $100 \div 25 = 4$. Multiply this by 3, because we're counting triplets of cats, and we get 12 cats.

- The answer is **13**:

    If 6 cats kill 6 rats in 6 minutes, then each cat kills a rat in 6 minutes. In 50 minutes, one cat can kill 8 rats: $50 \div 6 = 8\frac{1}{3}$. Note that the cat will have 2 minutes to spare, since $8 \times 6 = 48$ and you can't kill one-third of a rat. Theoretically, $100 \div 8 = 12\frac{1}{2}$ cats are needed to kill 100 rats, but since there are no "half-cats", the answer in practice is 13. With 13, all 100 rats are killed and if they so wished, the cats could even kill 4 more.

- The answer is **14**:

    If 6 cats kill 6 rats in 6 minutes, then 2 cats kill a rat in 3 minutes. In 50 minutes, every group of 2 cats can kill 16 rats with 2 minutes to spare, because $50 \div 3 = 16\frac{2}{3}$. To kill

100 rats in 50 minutes, 7 pairs of cats are needed because $100 \div 16 = 6\frac{1}{4}$ and there are no "quarter-cats", so more than 6 pairs are needed. Multiplying this by 2 — two cats a pair — and we get the required number, 14 cats. In fact, if they are really hungry they can kill 112 cats in the allotted time.

So, it's the cats' working procedure that determines the solution. If all the cats pounce together on each rat one at a time, or if they split into groups of 3 cats per rat, then 12 cats are needed. If each cat works individually, 13 cats is the answer. If their strategy is to work in pairs — perhaps one of them pins the rat down while the other pounces — the correct solution is 14.

The History of the Puzzle and Related Topics

Lewis Carroll is one of my childhood heroes. I spent hours reading *Alice in Wonderland, Through the Looking Glass* and *The Hunting of the Snark*. Little did I know then that I was "reading" math. Lewis Carroll is the pseudonym of the British mathematician Charles Lutwidge Dodgson. Born in Cheshire in 1832, the son of the Archdeacon of Richmond, he excelled in math, eventually earning him a position as lecturer at Oxford University. But Carroll did not make his fame through math. Rather, it was his literary genius, combined with his impeccable command of nonsense and social status as a profound entertainer for children and adults alike, that made him famous. Carroll used to perform magic tricks, games and displayed an unusual collection of bizarre inventions in front of social acquaintances.

Math, and in particular logic, made its way into Carroll's *Alice* books. These were first related as stories to the children of Henry Liddell, the dean of Christ Church, Oxford, whom Carroll used to take on short boating trips on the Thames river. Liddell's middle daughter Alice, whom the wonderland Alice is modeled after, was the one who persuaded Carroll to write down his stories. Alice received an honorary doctorate degree from Columbia University at the age of 80 for "awaking with her girlhood's charm the ingenious fancy of a mathematician familiar with imaginary quantities, stirring him to reveal his complete understanding of the heart of a child." This was in 1932. Such a sentence would probably not be at all appropriate these days, but Alice Liddell had many virtues and achievements of her own right. The Columbia quote also touched on the disturbing reports regarding Lewis Carroll's attraction to young girls. Jenny Wolf has published an excellent article in the *Smithsonian Magazine* regarding Lewis Carroll's "shifting reputation".

The popular math and science writer, Martin Gardner, wrote *The Annotated Alice*, a fantastic literary work uncovering, among other things, the math underlying many of the episodes in the *Alice* books. One of the things Gardner alludes to is Carroll's "pillow problems", puzzles that he tried to solve while lying in bed trying to cope with his insomnia.

> *"Only it is so very lonely here!" Alice said in a melancholy voice; and at the thought of her loneliness two large tears came rolling down her cheeks. "Oh, don't go on like that!" cried the poor Queen, wringing her hands in despair. "Consider what a great girl you are. Consider what a long way you've come today. Consider what o'clock it is. Consider anything, only don't cry!" Alice could not help laughing at this, even in the midst of her tears. "Can you keep from crying by considering things?" she asked. "That's the way it's done," the Queen said with great decision: "Nobody can do two things at once, you know."*
>
> *(Through the Looking Glass)*

Martin Gardner commented on this: "Carroll practiced the White Queen's advice. In his introduction to *Pillow Problems* he speaks of working mathematical problems in his head at night, during wakeful

Fig. 1.1   Lewis Carroll, 1863 photograph by Oscar G. Rejlander

hours, as a kind of mental work-therapy to prevent less wholesome thoughts from tormenting him."

Carroll's compendium of 72 pillow problems was first published in a book, *Curiosa Mathematica: Pillow-problems, thought out during sleepless nights* in 1893, five years before his death. Many of these puzzles are ingenious and some, notoriously difficult. Lewis Carroll published many more puzzles throughout his life, in various publications, and painstaking efforts have been made throughout the years, to locate and republish them. One such puzzle is our Cats and Rats puzzle which was published in 1879 in a puzzle column in *The Monthly Packet*, a magazine for girls, edited by Charlotte Yonge, but it was not Carroll who published it! Robin Wilson and Amirouche Moktefi in their recently published monumental work, *The Mathematical World of Charles L. Dodgson (Lewis Carroll)*, suggest that Carroll came across the puzzle and disliked it. Martin Gardner gives the details in his book, *The Universe in a Handkerchief*. Apparently, these kind of puzzles — *if a does b things in c minutes, how many a are needed to do d things in e minutes?* — were very popular and, it was suggested, have a unique solution, similar to the way rate problems were solved, rendering the "proper" solution of 12 cats. Lewis Carroll disliked the absurdity of the problem, so he published in the February 1880 issue of the magazine his "full" solution — those that we gave above — along with a puzzle of his own, mocking the original:

> *"If a cat can kill a rat in a minute, how long would it be killing 60,000 rats? Ah! How long indeed! My private opinion is that the rats would kill the cat."*

No recreational math book is complete without at least one Lewis Carroll reference. Why did he become so famous, both within the math community and the general public? I would argue that it was his mastery of math *and* the arts, especially writing and performing, coupled with the dramatic expansion of pastime opportunities, popular novels and plays of the Victorian era, that brought him his fame. Carroll was arguably the first person to use math for entertainment to the delight of many of the Victorian salons.

Generalizations of the Puzzle

Our minds have been trained to expect one answer for one puzzle. In one of the online puzzle courses that I run, one participant even protested that having more than one correct answer is "against the rules"! Since we make up the rules, there is of course no reason at all to restrict the number of solutions. In fact, in many real-world situations there can be quite a few solutions to a single puzzle.

Why and when do we get more than one answer to these and other puzzles? The crux of the matter is this: when some of the information is missing, there *may* be more than one solution to a puzzle. The missing information in the Cats and Rats puzzle is the cats' methodology — we do not know the cats' method of rat hunting. If they all pounce on each rat separately, or if three cats tackle each rat, then 12 cats are sufficient. 13 cats will solve the puzzle if each cat hunts down one rat, and if two cats pounce on one poor rat, then 14 cats are required.

What other puzzles can be classified as "missing information" puzzles? Here is another example — this time a "real-life" puzzle. Suppose you want to find the shortest distance between the corner of 5th Avenue and West 32nd Street in Manhattan to the corner of Madison Avenue and East 34th Street. For simplicity we take the length of an East–West street as one unit, and that of a North–South street as ½ unit because they are approximately a half length of the East–West streets as can be seen in Fig. 1.2 (this approximation is not far from the truth).

Most people who are in a "school geometry" frame of mind will use the Pythagorean theorem to calculate the distance and argue

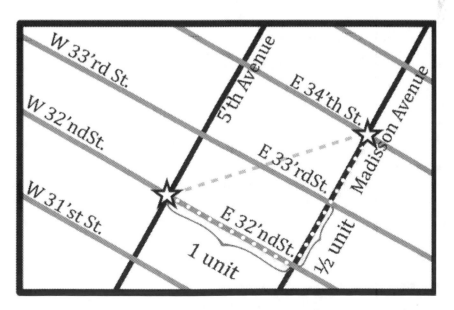

Fig. 1.2   The Manhattan route puzzle

for a shortest route of "the square root of 2" units. Using math terminology, this is the *hypotenuse* of an imaginary *right-angled triangle* whose *catheti (legs)* are:

- along 32nd Street between 5th Avenue and Madison Avenue = 1 unit long.

- along Madison Avenue between East 32nd Street and East 34th Street = two ½ units = 1 unit long.

According to the theorem, the length of the hypotenuse is $\sqrt{1^2 + 1^2} = \sqrt{2} = 1.414...$, nearly 1½ units long. This distance is shown as the dashed line in Fig. 1.2.

If you happen to be a pedestrian or a taxi driver for that matter, you're most likely to look for the shortest, *most practical* way to drive or walk between the two points. Since you don't walk or drive through buildings, you might argue that there are three "shortest routes", all three, 2 units long. The dotted lines in Fig. 1.2 show one of these three routes — going a block east on 32nd Street and then two blocks north on Madison Avenue. The "taxicab" calculation (as it is typically called) is straightforward. We just sum the street lengths of the different routes between the two points: 1 + ½ + ½ = 2.

So this puzzle has two "correct" answers: 2 or $\sqrt{2}$. The missing information in the puzzle is *topological.* Very broadly speaking, topology is the study of the shape of things. What is the topology of the puzzle? We just don't know! If we regard Manhattan as just one flat slab without streets or buildings, the topology is *planar* and we use *planar geometry* — the kind of geometry usually taught in school to calculate "as the crow flies" distances. But for a pedestrian or taxi driver, the problem should probably be solved within the framework of what's known as *taxicab geometry.*

Planar geometry has been known for millennia. As its name suggests, planar geometry problems occur in two-dimensional space. A typical, unambiguous, planar geometry "real-life" problem could be to calculate the shortest distance when driving from one spot to another on a flat, empty slab of asphalt — perhaps an unmarked parking lot. The answer would be a straight line. In planar geometry there is only one straight line between two points, so there is only one "shortest" route for this problem.

Planar geometry itself is part of a larger geometrical system known as *Euclidean geometry,* that includes three-dimensional solid geometry as well. This system was first described in *The Elements*, a series of 13 math textbooks, compiled by the 3rd century BCE Greek mathematician, Euclid of Alexandria. *The Elements* was the first rigorous attempt to lay the foundations of mathematical proof, containing much of the geometry and number theory known to man. Beginning with a set of five *axioms,* self-evident, unproved assumptions, the books devise a huge number of *theorems,* conclusions that arise from, and can be proved by these "accepted" truths. *The Elements* was so thorough and popular, that it was used as a textbook, in some places, even up to the 20th century! Over a thousand editions of *The Elements* have been published and it has been translated into many different languages.

The five planar geometry axioms, the Euclidean axioms or postulates, are usually taught in school. *Wolfram's Mathworld* defines these as:

1. A straight-line segment can be drawn joining any two points.

2. Any straight-line segment can be extended indefinitely in a straight line.

3. Given any straight-line segment, a circle can be drawn having the segment as radius and one endpoint as center.

4. All right angles are congruent.

5. If two lines are drawn which intersect a third in such a way that the sum of the inner angles on one side is less than two right angles, then the two lines inevitably must intersect each other on that side if extended far enough.

The fifth axiom is equivalent to what is known as the parallel postulate, and since it is not as "self-evident" as the other four axioms, mathematicians tried, time after time, to dispose of it by trying to deduce it from the other axioms. It was only in 1823 that both the Hungary mathematician, János Bolyai, and the Russian mathematician, Nikolai Lobachevsky, independently realized that not only is the parallel postulate essential for the completeness of Euclidean geometry, but that, by disposing of it, new "non-Euclidean geometric systems" arise. The 19th and 20th centuries witnessed the birth of elliptical and hyperbolic geometries — and many more.

Taxicab geometry was introduced by the 19th century Russian-born German mathematician, Hermann Minkowski, and was further developed into a system complete with its own axioms by Donald R. Byrkit in 1971. Distances in taxicab geometry are non-Euclidean. They are calculated only along horizontal and vertical segments on a Cartesian grid. Interestingly, the popular navigation app Waze™ shows both the planar distance, and the taxicab distance between two points of interest.

So far, we have seen two puzzles which give rise to different solutions, because some of the information in the puzzles is missing. This gives rise to the question of whether *all* puzzles and math problems are "missing information" puzzles that can be solved differently depending on how they are interpreted. We might be tempted to include some instances of algebraic equations. Quadratic equations, for example, have two solutions, so one might suggest that they are also problems with more than one solution because of missing information. But even if at first glance we can't explain why, we can feel that this is not the same genre as Lewis Carroll's Cats and Rats puzzle, or the Manhattan distance puzzle. In quadratic equations and other problems in algebra, everything is perfectly defined. The number of solutions is a manifestation of the equations themselves, independent of outside circumstances or interpretations. Conversely, "missing information" puzzles have different solutions precisely because of their interpretation.

Here are two other puzzles that might be considered "missing information" puzzles. You may have heard of them — they're quite popular — but their origins are not known.

- There are three on-off switches outside an attic, one of them controls an incandescent light bulb in the attic, but you don't know which one. You are allowed only one visit to the attic. How do you find the correct switch?

- Draw four straight lines that go through the middle of all the points on the following grid (Fig. 1.3) without lifting your pencil.

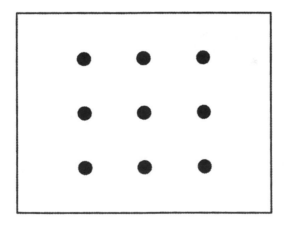

Fig. 1.3   The nine dots puzzle

In the earlier puzzles in this chapter at least one solution easily pops to mind, so much so, that you don't even question whether you have understood the puzzle correctly. Contrariwise, in these two puzzles, it seems at first as though there is no solution at all! The information that is given seems just not enough to solve the puzzle. A lot of the time, when confronted with puzzles like these, people realize that information must be missing, so they tend to ask questions to get that information. In the attic light bulb puzzle, a question might be: can you see the attic from where the switches are? In the nine dots puzzle, people might ask: what do you mean by straight lines? However, in cases like these, the missing information is not usually what it seems and you need to look "outside the problem" for at least one solution. These kinds of puzzles are also known as puzzles with solutions that are "out-of-the-box".

The attic light bulb puzzle has one popular solution. Turn on two of the switches, wait a few minutes, turn off one of the two switches and then go into the attic. If the light is on, then the switch that is currently in the "ON" position controls the light in the attic. If the light is off but the light bulb is warm, then the switch that was turned "ON" and then "OFF" controls the light. If the light bulb is cold, the remaining switch is the solution.

As often happens in "missing information" puzzles, I have also come across some more bizarre solutions. Mathematician and puzzler, Peter Winkler, suggests that for those who can't reach the light bulb but have a lot of time, to turn on one switch, wait two months, turn it off and turn on another switch and then visit the attic. The first switch controls the light if the light bulb is burnt out. Of course, this solution assumes that incandescent light bulbs are still in use — in many places nowadays they have been replaced by eco-friendly alternatives. The solution also assumes that you can visibly see when a light bulb is burnt out.

The solution to the nine dots puzzle literally *does* require out-of-the-box thinking, as can be seen in Fig. 1.4.

What *was* the missing information in these two puzzles? In the attic light bulb puzzle, it was the fact that you are allowed to use physics (i.e. feel the light bulb). In the nine dots puzzle it was that you are not confined to the nine-dot grid. But are these pieces of information missing? Only in the minds of the solver — there is nothing in the puzzles to suggest that!

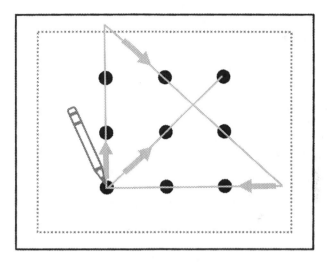

Fig. 1.4   Solution to the nine dots puzzle

Recap

Lewis Carroll's Cats and Rats puzzle is famous for having more than one solution — not because this is mathematically imperative, but because the puzzle can be interpreted in different ways due to some "missing information". There are many types of "missing information" puzzles. We saw three examples. In the first example, the Cats and Rats puzzle, the missing information was *methodological* — what was the cats' method for killing the rats? In the second example, the Manhattan route puzzle, the missing information was *topological* — what was the geometrical layout of the puzzle? In the attic light bulb and nine dots puzzles, the missing information was completely outside the *assumed* sphere of the puzzle; you are not confined to the puzzle's grid or border (the nine dots puzzle) or to math (the attic light bulb puzzle).

Along the way, we briefly introduced a few concepts that you might want to remember. Here are some short, informal definitions:

- *Out-of-the-box puzzles* — puzzles where the solution can be found outside the perceived boundaries of the puzzle.

- *Pillow problem* — a problem or puzzle that you "sleep on". The term was coined by the British mathematician, Charles Lutwidge Dodgson (aka Lewis Carroll), the author of the *Alice in Wonderland* books.

- *Planar geometry* — the study of shapes in the plane.

- *Pythagorean theorem* — the square of the lengths of the legs in a right triangle is equal to the square of the hypotenuse.

- *Taxicab geometry* — the study of shapes on a grid.

- *Topology* — the study of the properties that are preserved through certain, continuous deformations of objects, such as bending, stretching and twisting, but not puncturing or cutting.

Challenge Yourself!

1. This is one of Lewis Carroll's most famous pillow problems. A bag contains one counter, known to be either white or black (with equal probability). A white counter is put in, the bag shaken, and a counter drawn out, which proves to be white. What is now the chance of drawing a white counter?

2. British puzzler, Henry Dudeney, proposed the following "missing information" puzzle: if you add the square of Tom's age to Mary's age, the sum is 62; but if you add the square of Mary's age to Tom's age, the result is 176. What are the ages of Tom and Mary? Apart from solving the puzzle, can you tell what the missing information is, and how you can get around it?

3. If you add my age to yours, you'll get 66. My age is your age in reverse. How old are we?

4. There are two different values of π — the ratio between the circumference of a circle and its diameter — depending on whether we are discussing a planar geometry circle or a taxicab circle. Find both of them!

5. Place the numbers 1–6 in the circles on the "magic" triangle in Fig. 1.5, so that the sum along each of the edges is the same. There are four possible solutions. What constraint would you add so that the solution is unique?

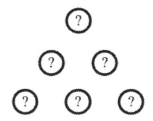

Fig. 1.5   Magic triangle puzzle

6. It takes six minutes to boil an egg. How long does it take to boil three eggs?

Solutions

1. Two thirds. There are three possible cases, in two of them the counter in the bag is white:
   (a) The initial counter remained in the bag and is white, and the one removed was the white counter that was put in.
   (b) The initial counter remained in the bag and is black, and the one removed was the white counter that was put in.
   (c) The white counter that was put in is now the one in the bag and the counter removed was white.

2. Tom's age is 7, Mary's age is 13. One way of solving this is by using algebra — but that does get difficult. An easier way is by understanding the missing information; there are two things missing that we can infer:

   (a) We can assume that the ages will be whole numbers

   (b) We can assume that the ages will be between 0 and 120 (more or less ...)

   This makes things much easier. Tom's age has to be smaller than 8 because $8^2 = 64$ which is larger than the sum of the square of Tom's age plus Mary's age (given as 62).

   Since the square of Mary's age plus Tom's age is 176, Mary's age has to be at least 8 (otherwise Tom is older than 112) but less than 14 ($14^2 = 196$). Trial and error on the four possible ages of Mary gives the answer.

3. There are four possible solutions to this. Our ages can be 6 and 60, 51 and 15, 33 and 33, or 42 and 24.

4. The typical (planar) value of π — the ratio between the circumference of a circle and its diameter — is 3.14159... .

   To find the value of taxicab π, we first need to define a taxicab circle, and then divide its circumference by its diameter. A circle is the collection of points at an equidistance from its center. An example can help. Figure 1.6 shows a taxicab circle with a radius of three units. Funnily enough, it's the shape of a diamond! The circumference — the distance along the perimeter of the circle until returning to the initial point is 24 units. Dividing that by the diameter (twice the radius = 6 units), gives $\pi = 4$. You can see for yourself that changing the radius will change the circumference proportionally, so the value of $\pi = 4$ is indeed constant.

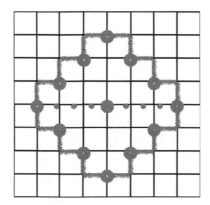

Fig. 1.6   A taxicab circle

5. Figure 1.7 shows four possible solutions, up to symmetry. A constraint might be — find the solution that has the sides with the lowest/highest sum.

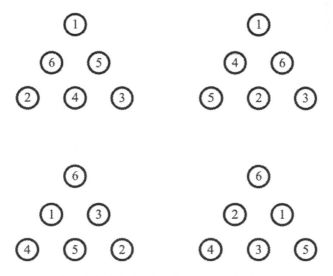

Fig. 1.7   Magic triangle puzzle solutions

6. This is a classic "missing information" puzzle. It all depends on how many saucepans and burner rings you have! If you have one saucepan that holds all three eggs or three saucepans that hold one egg each and three burner rings, or two saucepans (one that holds one egg and the other, two eggs) and two burner rings, then you can boil all three eggs in six minutes. If you have two saucepans that hold only one egg each and two burner rings, or at least one saucepan that holds only two eggs and one burner ring, then it will take you twelve minutes. (Oded Margalit pointed out to me that this is assuming that each egg has to boil six minutes continuously. If we can remove and replace eggs as we like, then we can boil two eggs on two burners in just nine minutes — boil eggs 1 and 2 for three minutes, then boil eggs 1 and 3 for three minutes, and finally, boil eggs 2 and 3 for three minutes.) If you have just one saucepan that holds one egg and one burner ring, then you'll have to wait a whole eighteen minutes to boil all three eggs.

Bibliography and Further Reading

Byrkit, D. R. (1964). Taxicab geometry—a non-Euclidean geometry of lattice points, *The Mathematics Teacher*, **64**(5), pp. 418–422.

Carroll, L., Gardner, M., Burstein, M., & Tenniel, J. (2015). *The Annotated Alice: Alice's Adventures in Wonderland & Through the Looking-glass*, 150th Anniversary Deluxe Edition. W. W. Norton & Company, USA.

Gardner, M. (1996). *The Universe in a Handkerchief: Lewis Carroll's Mathematical Recreations, Games, Puzzles, and Word Plays*. Copernicus, New York.

Levitin, A., & Levitin, M. (2011). *Algorithmic Puzzles*. Oxford University Press, Oxford.

Weisstein, E. W. (n.d.), Euclid's Postulates, from *MathWorld*—A Wolfram Web Resource. http://mathworld.wolfram.com/EuclidsPostulates.html

Weisstein, E. W. (n.d.), Topology, from *MathWorld*—A Wolfram Web Resource. https://mathworld.wolfram.com/Topology.html

Wilson, R., & Moktefi, A. (Eds.) (2019). *The Mathematical World of Charles L. Dodgson (Lewis Carroll)*. Oxford University Press, Oxford.

Winkler, P. (2003). *Mathematical Puzzles: A Connoisseur's Collection*. CRC Press, USA.

Wolf, J. (2010). Lewis Carroll's Shifting Reputation, *Smithsonian Magazine*. https://www.smithsonianmag.com/arts-culture/lewis-carrolls-shifting-reputation-9432378/

chapter

2

Retrolife Puzzles

Introduction

Many people and puzzles have been influenced by one man, namely Martin Gardner, arguably mathematics' most prolific writer and communicator. Gardner earned his fame from his monumental mathematical games columns, which he wrote for *Scientific American* between the years 1957 to 1981, and from dozens of recreational math and popular science books that he wrote. One of his top articles that had a tremendous impact on the math and computer science communities, investigated celebrated mathematician John Horton Conway's so-called Game of Life. Retrolife is a puzzle based on the game, which truly has a very long tale.

The Puzzle

A word of caution before the puzzle. The puzzle rules seem quite bizarre, but bear with them, and you'll really enjoy the chapter. Trust me!

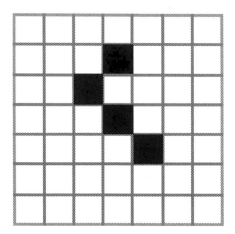

Fig. 2.1   Retrolife puzzle.
Target number of marked white squares: 9

A board (Fig. 2.1) consists of black and white squares, where the black squares are arranged in a pattern. Every square on the board, whether black or white, is (obviously) surrounded by 8 neighboring squares in the directions: North, North–East, East, South–East, South, South–West, West, North–West (except those on the border, so assume the board is infinite).

Mark (with an X) **9 white squares** *while obeying the following rules:*

- *Each black square must be surrounded by exactly 3 marked white squares.*

- *Each marked white square may be surrounded by any number of other marked white squares (including 0), except 2 or 3 other marked white squares.*

- *Each unmarked white square may be surrounded by any number of marked white squares (including 0), except 3 other marked white squares.*

Where to Start?

Sometimes, understanding the rules of puzzles is not easy. When you take a first look at the puzzle above, you probably ask yourself, why 9? However, when you try to solve the puzzle, you'll find that in fact 9 is necessarily the minimum number needed to ensure that all the rules are observed.

We sometimes jump into difficult puzzles without going through easier ones. This is often the case with puzzles like Sudoko, Rush Hour (Traffic Jam), Area Mazes and so on. But practice does indeed make perfect, and solving easier puzzles pays off in the long run. Figure 2.2 shows an easier puzzle and its solution. The rules are the same. In this case, only 7 marked white squares are needed. Try and solve the puzzle first, before peeking at the solution. The solution is not unique; there are, in fact, 20 different solutions with 7 marked white squares, so if you solved the puzzle in a different way, that might still be alright. I would recommend though, making absolutely sure that the rules have not been violated. Later on, when we get down to the origins of this puzzle, I'll show you an easy way to check your solution.

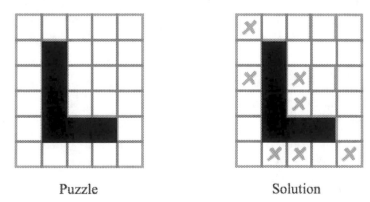

Puzzle   Solution

Fig. 2.2   Retrolife puzzle and solution.
Target number of marked white squares: 7

Take a look at the puzzle and its solution to convince yourself that every black square has exactly 3 marks (X's) in its neighboring squares. You can find unmarked white squares with 0, 1 and 2 marked white neighbors, but none with exactly 3 marked white neighbors. This is the only restriction on the unmarked white squares. Note the restriction for the black squares (that they have exactly 3 marked white neighbors) is the exact opposite of the restriction for the unmarked white squares (that they do not have exactly 3 marked white neighbors).

There is one more restriction: a marked white square must not have 2 or 3 marked white neighbors. In the puzzle shown in Fig. 2.2 every marked white square has either 0 or 1 marked white neighbors; none has 2 or 3.

Solving the Puzzle

Here is the unique solution to the puzzle, confirmed using a computer program written by Neil Bickford.

Figure 2.3 shows the unique solution to the puzzle. The difficulty of this puzzle is one notch up from the example above (Fig. 2.2). Note the lone X in the middle of the first column. Although it is not adjacent to any black cell, its presence is unavoidable. Without it, the white square in the third row of the second column would have three neighboring X's, breaking one of the rules. Some of the puzzles at the end of this chapter require some elaborate structures to avoid breaking the rules. Figure 2.4 gives an example of such a structure.

You might wonder why the target number, 19, is so large, given that there are only six black squares. Surprisingly, 19 is the *minimum* number of marks that solve this puzzle, and this can be proved using logical reasoning. Take a look at the center black square. It has only three surrounding white squares, so marking them is inevitable. Having done this, we notice that the lower marked white square has

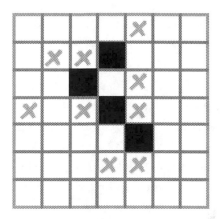

Fig. 2.3 Retrolife puzzle and solution.
Target number of marked white squares: 9

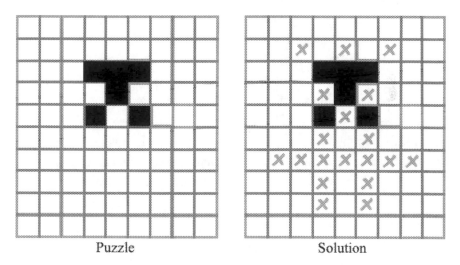

Puzzle   Solution

Fig. 2.4 Retrolife puzzle and solution.
Target number of marked white squares: 19

two marked squares, violating the second condition. So, we have to mark some of *its* surrounding white squares as well. This, in turn, creates some more problematic squares, resulting in the leg-like structure underneath the pattern.

The History of the Puzzle and Related Topics

I invented Retrolife in 2005, as a puzzle for kids in the Davidson Institute's recreational math club, then known as "Math-by-Mail", while we were learning about cellular automata (mathematical models for dynamical systems that change as a function of time). Retrolife has roots in one such automaton, the Game of Life, invented in the 1960's by the famous British–American mathematician John Horton Conway, who passed away due to COVID-19 in March, 2020. Apart from being a rigorous mathematician, he was a great *recreational* mathematician and math educator, who inspired many young kids with his love for math.

John Conway has been described as a cross between Mick Jagger, Salvador Dali and Archimedes. Born in Liverpool, England in 1937, he quickly became one of the most profound and beloved mathematicians of the 20th century. As a teenager, he described himself as a quiet, math-loving introvert, but later on, as a student at Cambridge University, and for the rest of his life, he shed his former personality to become a "partyholic hippy", alongside his superb career in math. Working first at Cambridge and later on at Princeton University in the United States, he made many contributions to game theory, group theory, topology, geometry, algebra and even theoretical physics, where together with Canadian mathematician Simon B. Kochen, he formulated and proved the free will theorem.

His other achievements include his phenomenal work on mathematical games, that he published in a four-book series, *Winning Ways for your Mathematical Plays*, written together with Richard Guy and Elwyn Berlekamp, both of them reputed mathematicians in their own right. Coincidentally, all three authors died within a year of each other. Conway also wrote *On Numbers and Games*, where he explained his discovery of surreal numbers. He was known for his ability to connect seemingly disparate areas of math, like geometry and group theory, resulting in his discovery of what's known as the "monster group".

Conway sported a great sense of humor, and was a very friendly, outgoing person, but his personal life was full of ups and downs. He had bouts of depression and survived both a heart attack and a stroke. One of his favorite pastimes was math education and games, so it wasn't very surprising that he was a formidable chess and backgammon player, usually making up new rules to make the games more interesting. In fact, it was while he was playing with some game counters on the square-tiled floor of his office, that he invented the Game of Life, for which he is probably best known.

The Game of Life is not really a game at all. It is a fundamental mathematical model for the evolution of systems. The system's "universe" is just a (usually infinite) grid, with each square in the grid in a state of either "ON" or "OFF". Known as "live" and "dead" cells, they play the part of the elements of the system. If the system is modeling epidemiology, for example, the cells might be live and dead viruses. When modeling the spread of forest fires, three states are needed: "UP", "DOWN" and "CENTER" describing unburnt, burnt and burning trees respectively, but the idea — modeling systems through a discrete number of states — is the same. The "game", better described as a simulation, starts with an initial configuration that the "player" sets up, where some of the cells are alive, and the rest, usually the majority, dead. Conventionally, live cells are the black squares on the board and dead cells are the white squares. The initial setup, or state, of the board is static. Conway derived a set of rules that are applied to the state of the board to mimic evolution. The Game of Life rules dictate for each cell on the board whether it will die (if it was previously alive) or be born (if it was previously dead), depending on what's going in its eight neighboring cells. When applied once, these rules determine what happens to each of the cells in one time step, known as a *generation*; these rules determine which live cells die and which are born. A solitary live cell

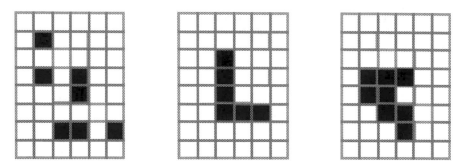

Fig. 2.5   Three consecutive generations in the Game of Life, from left to right.

that has either no living neighbors or just one, will become a dead cell in the next time step. A live cell that has more than three live neighbors, will also die, mimicking overcrowding (note the similarity to the behavior of viruses). The "birth" of a live cell, replacing a dead cell, occurs when the dead cell has exactly three live neighbors. In all other cases, nothing changes. Figure 2.5 shows three consecutive generations in the Game of Life.

When iterated for *many* time steps the evolution of the whole system as a function of time can be seen, sometimes resulting in some very interesting and unexpected outcomes. It took Conway a lot of trial and error to come up with the specific rules that make the game interesting, and it turned out that he did a great job, with the Game of Life being able to model even a fully-functional computer (known in professional jargon as being Turing complete). Since its publication in 1970 through a series of articles written by math popularizer and writer, Martin Gardner, in *Scientific American*, interest in the Game of Life surged and it became the benchmark for all computer simulations from earthquake modeling to the SIMS™ computer games. You will not find a single computer scientist or mathematician who hasn't heard of it. It also brought fame to Conway, who became widely known outside the math community, and continued to regularly share his passions and discoveries with the public until his untimely death. Yet, he often complained that he "hated" the Game of Life, mainly because it took away the focus from his other mathematical achievements.

Retrolife is the inverse of the Game of Life, with some additional restrictions. In the Retrolife puzzle, you are asked to find the precursor state (i.e. one generation before), given a certain state in the Game of Life. For example, what is the precursor state if you are told that all the live cells in the given state were dead in the precursor state, and that the precursor state had the minimum number of live cells (we call this the target number)? You can see that the center image in Fig. 2.5 is the Retrolife puzzle in Fig. 2.2, and the image on the left in Fig. 2.5 is its precursor state, where there are marked squares in the solution in Fig. 2.2. This makes it easier to check Retrolife solutions — just roll them forward in the Game of Life and see that you reproduce the puzzle.

Searching for predecessors in the Game of Life is not new. Two mathematicians, Edward Moore and John Myhill, proved that in many cellular automata, there exist states without predecessors, that is, states that cannot be born — generation 0 if you like. Accordingly, John Tukey, an American mathematician who studied cellular automata, called these generation 0 states "Gardens of Eden". The Game of Life falls under Moore and Myhill's categories, so from its inception, Conway and others knew there should be Garden of Eden states, but it turned out they were notoriously hard to find, even on ultra-fast computers. MIT professor Robert Banks and fellow workers found the first Garden of Eden in the Game of Life in 1971 which consisted of 226 living cells, and to date, only 25 Gardens of Eden have been found. (Theoretically, once a Garden of Eden pattern is found, one could scatter single cells far away from the pattern to create a hypothetical, infinite number of states from each Garden of Eden, but naturally, these are not considered "connected", and are not counted.)

The first Game of Life predecessor search as a puzzle, as far as I know (and thanks to Tim Chow for bringing this to my attention), is the traditional MIT mystery hunt, an annual puzzle hunt competition for students. David Reiley incorporated a predecessor search for a given Game of Life state in the 1996 competition. The puzzle was to find the predecessor state for what looked like a scattering of live cells to reveal a secret message. The conditions that we added — that the live cells that create the pattern on the puzzle board are dead in its predecessor state, and that there is a target number of live cells to find — enabled Retrolife to detach its immediate connection to the Game of Life and serve as a puzzle in its own right, so puzzle enthusiasts don't necessarily have to know about the Game of Life in order to solve it.

Fig. 2.6   John Conway (right) and the author at a conference where Retrolife was presented

Generalizations of the Puzzle

There are many ways to generalize the idea of Retrolife. First, we can change the rules. Here are the original conditions again:

- *Each black square must be surrounded by exactly 3 marked white squares.*

- *Each marked white square may be surrounded by any number of other marked white squares (including 0), except 2 or 3 other marked white squares.*

- *Each unmarked white square may be surrounded by any number of marked white squares (including 0), except 3 other marked white squares.*

We can change the third condition, for example, to:

- *Each unmarked white square may be surrounded by any number of marked white squares (including 0), except **2** other marked white squares.*

   Funnily enough, even this very slight change makes some Retrolife puzzles much harder to solve. Figure 2.7 shows two solutions to a Retrolife puzzle. The image on the left shows the solution to the original version, while that on the right shows the modified condition, where unmarked white squares cannot have exactly 2 other marked white neighbors.

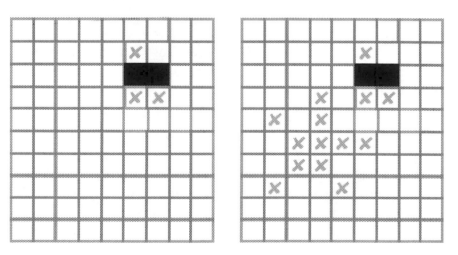

Fig. 2.7   Retrolife puzzle and solution.
Left: regular conditions; right: modified conditions

Another possibility is to change the *shape* of the board. For example, Fig. 2.8 shows one example of Retrolife on a hexagonal board, and its solution (using the original conditions).

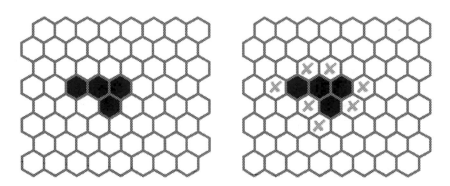

Fig. 2.8   Retrolife puzzle on a hexagonal board and its solution

We can also transform Retrolife from a one-person puzzle to a two-or-more player game. This can be done in a few ways. The simplest is to randomly choose a pattern of black squares for the puzzle, perhaps from a given set, and the player who can solve it with the minimum number of marks, wins. One step up is for each player to challenge the other with a pattern, where the number of black squares on the board is mutually decided upon. For example, at the outset of the game the players decide upon a number, say four, and then each player creates a Retrolife pattern with four black squares that the other players have to solve, perhaps within an allotted time. The winner is the one who can solve the puzzle given to him, first. If the pattern is such that there is no solution, a player declares "no solution", and wins, unless he is wrong and the other player finds one, in which case he is the winner.

Two players (or more) can also play a game using an initially selected pattern on *one* board, taking turns to mark one of the white squares. A player loses if he marks a square that renders the solution impossible (for example, marking a white square adjacent to a black square that already has three marked neighbors). The winner is the player who places the last mark.

The Game of Life was one of the first cellular automata to be discovered — or perhaps we should say, "created". Many more automata have since been developed. Computer scientist, theoretical physicist, author and CEO of Wolfram Research software company, Stephen Wolfram, is one of the pioneers of cellular automata computational modeling. Wolfram, who created the popular technical computing system, Mathematica, and its artificial intelligence counterpart Wolfram Alpha, published in 2002 a huge bestseller, *A New Kind of Science* to establish a fundamental postulate that cellular automata can serve as models for practically anything in real life, replacing the notion that mathematical equations in the natural sciences are the only "proper" tools. It is truly astounding that very complex behavior can arise from the simple rules of cellular automata.

Wolfram's favorite automaton, is a one-dimensional cellular automaton that he calls "Rule 30". Being one-dimensional means that instead of a grid we have just one line, that evolves in time. It is customary to write each time evolution of a 1-d cellular automaton as a new row, written underneath the previous generation. Thus, a grid is formed where we can view the entire time evolution of the initial state (the first row) at one glance. The grid is not to be confused with a single state on a 2-d automaton.

Each cell in a new generation (row) can be live (black for "ON") — or dead (white for "OFF") — we'll use colors henceforth. In "Rule 30" (and many other 1-d automata) this is determined according to the state of three cells in the previous generation — the one that is directly "above" the cell we are considering, and the cells to its immediate right and left. This is analogous to the conditions of the neighbors in the Game of Life, that collectively make up the rule for generating the next state on the board. Figure 2.9 shows the conditions in "Rule 30". There are eight

This color pattern of three cells at generation: t →
specifies **White** for the center cell directly below at generation: $t+1$ →

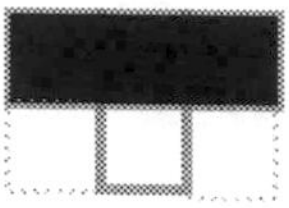

This color pattern of three cells at generation: t →
specifies **White** for the center cell directly below at generation: $t+1$ →

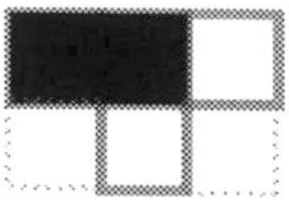

This color pattern of three cells at generation: t →
specifies **White** for the center cell directly below at generation: $t+1$ →

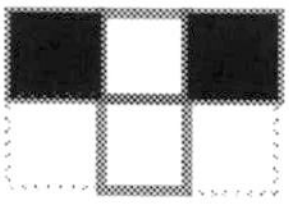

This color pattern of three cells at generation: t →
specifies **Black** for the center cell directly below at generation: $t+1$ →

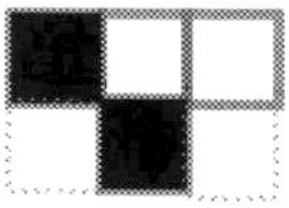

This color pattern of three cells at generation: t →
specifies **Black** for the center cell directly below at generation: $t+1$ →

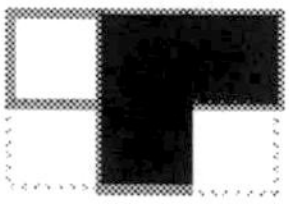

This color pattern of three cells at generation: t →
specifies **Black** for the center cell directly below at generation: $t+1$ →

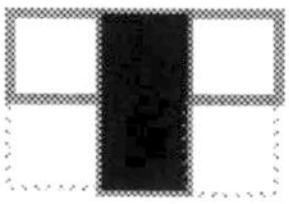

This color pattern of three cells at generation: t →
specifies **Black** for the center cell directly below at generation: $t+1$ →

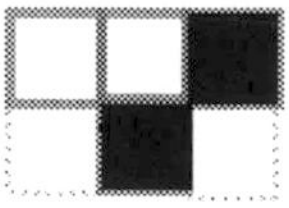

This color pattern of three cells at generation: t →
specifies **White** for the center cell directly below at generation: $t+1$ →

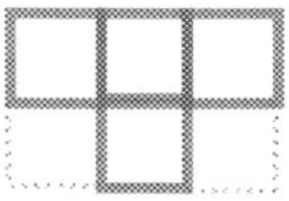

Fig. 2.9   Wolfram's "Rule 30"

conditions in all, because there are eight different color combinations for three cells in a row. Using B for black and W for white, these combinations are: BBB, BBW, BWB, BWW, WBB, WBW, WWB and WWW. The first condition in Fig. 2.9 means: if the three cells are all black, then the color of the center cell directly beneath them should be white, BBB → W. The second condition means: if the cells, from left to right, are black, black white, then the color of the center cell beneath them in the next generation has to be white, BBW → W. The remaining conditions are: BWB → W, BWW → B, WBB → B, WBW → B, WWB → B and WWW → W.

Figure 2.10 shows the evolution of a solitary black cell according to "Rule 30" for five time steps (generations). Each row in the grid is a new generation, and the color of each cell in every row is determined according to the three cells immediately above it.

For example, beginning from the left-hand side of the grid, the first three cells in the second row has three white cells directly above which, according to the conditions, dictate a white cell. (Technically, the leftmost and rightmost cells have only two cells above them, because the extreme left/right cell falls outside the grid, but these are considered white by definition.) Continuing along the row we reach a cell where the three cells above it are: white, white, black. According to the conditions of "Rule 30" this dictates a black cell. The next cell has white, black, white above it, dictating

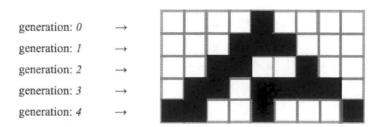

generation: *0* →
generation: *1* →
generation: *2* →
generation: *3* →
generation: *4* →

Fig. 2.10   First five generations of a single black square according to Wolfram's "Rule 30" a black cell. The next cell along has black, white, white in the row above it, meaning the cell has to be black. From here onwards, we're back to white cells, because all remaining cells have three white cells immediately above them.

The true beauty and complexity of "Rule 30" can be seen when it is iterated for a very large number of generations. Figure 2.11 shows how complex and beautiful this is for the initial condition of a solitary black cell.

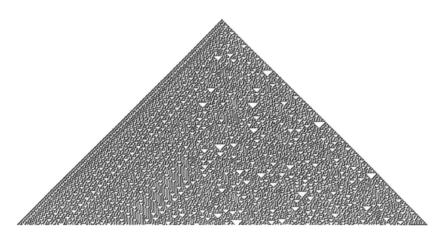

Fig. 2.11   Iteration of Wolfram's "Rule 30" for a large number of time steps

But why is this so important? It turns out that, just like the Game of Life, although stemming from very simple rules, the outcome is surprisingly unpredictable. Using mathematical terms, the evolution using "Rule 30" is both *aperiodic* and *chaotic*. *Aperiodic* means that there is no recurring pattern; the pattern does not recur at regular intervals. *Chaotic* means that the outcome — the pattern that can be seen after many iterations — is very sensitive to the initial conditions. Figure 2.12 shows two 100-step simulations using "Rule 30" on an initial row (this time the initial row is made of a "sprinkle" of live cells). Although having the same overall "look",

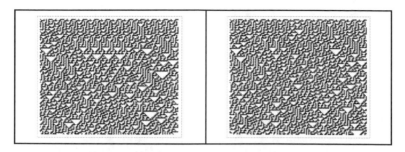

Fig. 2.12   Two simulations of Wolfram's "Rule 30". The first rows — the
initial conditions — of each simulation are identical except for one cell
that is white in the image on the left and black in the image on the right

the evolved patterns are very different, yet the first rows in each
evolution differ by just one cell that was chosen to be black in one
simulation and white in the other!

Aperiodic and chaotic simulations are extremely useful and
computationally interesting. They serve as models for many natural
phenomena like predicting the weather, and for generating random
sequences — something very important in modern cryptography
and cybersecurity.

Is there a one-dimensional version of Retrolife? Certainly! One
can give puzzle solvers a row and a rule and ask puzzle solvers to
find the previous generation. For "Rule 30", nothing more is needed,
because this is a *reversible* cellular automaton. Given generation 4
in Fig. 2.10, and the conditions of "Rule 30" (Fig. 2.9), solvers will
be able to find the previous generation (generation 3). The one-to-
one correspondence in this case, makes this inverse problem more
"rigorous" and less "puzzly". There is exactly one way to go forward
a generation, and, given the conditions and a state, there is exactly
one way to go backward and recover the previous generation.
This is what's known as a "reverse engineering" problem, that
can also be solved using advanced math such as linear algebra. But
not all one-dimensional cellular automata are reversible. Wolfram's

"Rule 90" for example has one way forward, but four ways backward. This means that, given just a row and "Rule 90", there are four possible predecessors. If we want to make this into a puzzle, we will need extra conditions, making it similar to the original Retrolife puzzle (recall that the Game of Life is not reversible).

Finally, a note on *how* Wolfram numbers his one-dimensional cellular automata rules, 30, 90, etc. Each rule is made up of eight conditions, all the combinations of three cells: BBB, BBW, BWB, BWW, WBB, WBW, WWB and WWW. Replacing B with the digit 1 and W with the digit 0, this translates to: 111, 110, 101, 100, 011, 010, 001 and 000. These are the *binary numbers* in countdown order from 7 to 0.

Binary numbers are numbers as they are represented in base 2. The number system that we use today is base 10 meaning that we denote numbers using ten digits: 0, 1, 2, 3, 4, 5, 6, 7, 8 and 9. To denote numbers larger than 9, we add a "tens" place and now all numbers between 0 and 99 can be written. For larger numbers we have the hundreds, thousands, etc. place, with each additional place adding a power of 10.

Suppose we only had two digits, 0 and 1. We could still write down all the numbers, but now the "tens" place would be the "two's" place, the "hundreds" would be the "four's" place and so on, with each additional place adding another power of 2. After the numbers 0 and 1, we write 10 instead of 2, because we are only allowed to use the two digits 0 and 1. We "tick over" to using the next place for larger numbers much quicker than the ten-digit case. When using binary, it is customary to use the word *bits* instead of digits. We say that there are ten digits, 0–9, but just two bits: 0 and 1. Table 2.1 shows the first 16 whole numbers in decimal and binary (base 2).

Let's take another look at the conditions of "Rule 30" in binary and in countdown order:

$111 \rightarrow 0, 110 \rightarrow 0, 101 \rightarrow 0, 100 \rightarrow 1, 011 \rightarrow 1,$
$010 \rightarrow 1, 001 \rightarrow 1$ and $000 \rightarrow 0$.

Now, let's take just the outcome — the bits written after each of the arrows that signals the color of the cell — in sequential order from 111 down to 000. We get 00011110, and this, in binary, is the number 30 — "Rule 30". This way, one number says it all! Wolfram constructed, studied and named all 256 possible three-neighbor rules, and wrote about them in his phenomenal book, *A New Kind of Science*, accompanying each rule with images of its long-term evolution given for some initial condition on the zeroth generation.

Table 2.1 The first 16 whole numbers in decimal and binary (base 2)

Number value (decimal)	Base 2
0	0
1	1
2	10
3	11
4	100
5	101
6	110
7	111
8	1000
9	1001
10	1010
11	1011
12	1100
13	1101
14	1110
15	1111
16	10000

Recap

In a Retrolife puzzle, given a pattern of black squares on a grid of white squares, you are to mark a target number of white squares that meet a few conditions: each black square must have exactly 3 marked neighbors; each marked white square may not have 2 or 3 marked neighbors; each unmarked white square may not have exactly 3 marked neighbors.

We analyzed Retrolife from different angles. We saw that the puzzle is strongly connected to two-dimensional cellular automata. In particular, it is a reverse engineering problem on a given state in John Conway's Game of Life, under certain conditions. We saw what other puzzles can be created if we changed the conditions, the board, or the number of players. We also learnt about Stephen Wolfram's set of one-dimensional cellular automata and their binary naming system. The evolution of an initial state in some cellular automata can result in some surprisingly complex structures, which is one of the reasons they are fundamental building blocks for simulating chaotic and aperiodic systems.

Here are some short, informal definitions of some of the interesting things we saw in this chapter:

- *Binary* — a base-2 counting system using only the digits 0 and 1, called *bits*.

- *Cellular automaton* — a system composed of some units, often called cells that can be in a finite number of states (i.e. "ON" and "OFF"), an initial condition and a set of rules that govern its evolution with time.

- *Game of Life* — John Conway's two-dimensional cellular automaton that is played on an infinite square grid, within which each cell "is born", "dies" or "survives" from generation to generation, according to conditions on its neighbors.

- *Retrolife* — a puzzle where you are required to mark certain white squares that surround a pattern of black squares on a white-squared grid according to given conditions.

- *Reverse engineering* — reproduction of the structure of a system, given its evolutionary state. Retrolife is a "reverse engineering" puzzle of a given generation in the Game of Life.

- *Stephen Wolfram's one-dimensional cellular automaton* — cellular automaton where each generation (time step) is an infinite row in a table and the evolution rules for each cell depend on three adjoining cells in its predecessor generation.

Retrolife is probably my favorite puzzle. I can spend hours solving these puzzles. Maybe it's the connection with the iconic Game of Life, or perhaps just the colorful character of John Conway. Whatever the case, once you get the hang of it, there's plenty of fun.

Challenge Yourself!

1. "Classic" Retrolife puzzles:
   In Figs. 2.13 to 2.16 mark (with an X) the given target number of white squares while obeying the following rules:

   - Each black square must be surrounded by exactly 3 marked white squares.

   - Each marked white square may be surrounded by any number of other marked white squares (including 0), except 2 or 3 other marked white squares.

   - Each unmarked white square may be surrounded by any number of marked white squares (including 0), except 3 other marked white squares.

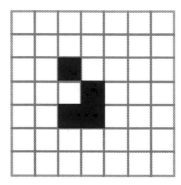

Fig. 2.13   Retrolife puzzle (a).
Target number of marked white squares: **6**

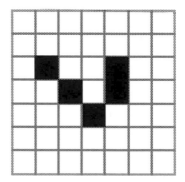

Fig. 2.14   Retrolife puzzle (b).
Target number of marked white squares: **9**

Fig. 2.15   Retrolife puzzle (c).
Target number of marked white squares: **13**

Fig. 2.16   Retrolife puzzle (d).
Target number of marked white squares: **19**

2. What is the predecessor of the following state in the Game of Life, given no conditions at all, using the minimum number of live cells?

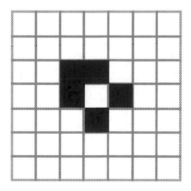

Fig. 2.17   Find the minimum predecessor, no extra conditions

3. A Retrolife puzzle with a different condition. Mark (with an *X*) **10** white squares while obeying the following rules:

   • Each black square must be surrounded by exactly 3 marked white squares.

- Each marked white square may be surrounded by any number of other marked white squares (including 0), except 2 or 3 other marked white squares.

- Each unmarked white square may be surrounded by any number of marked white squares (including 0), except <u>2</u> other marked white squares.

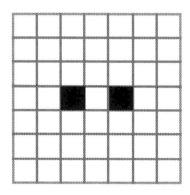

Fig. 2.18   A Retrolife puzzle with a different condition

4. Hexagonal Retrolife puzzle. Mark (with an X) **the minimum number of** white hexagons needed to obey the following rules:

   - Each black hexagon must be surrounded by exactly 3 marked white hexagons.

   - Each marked white hexagon may be surrounded by any number of other marked white hexagons (including 0), except 2 or 3 other marked white hexagons.

   - Each unmarked white hexagon may be surrounded by any number of marked white hexagons (including 0), except 3 other marked white hexagons.

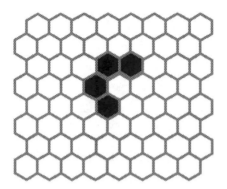

Fig. 2.19   A hexagonal Retrolife puzzle

Solutions

1. "Classic" Retrolife puzzles:

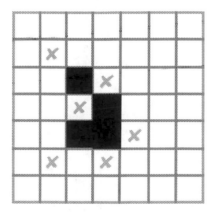

Fig. 2.20   Solution to Retrolife puzzle (a).
Target number of marked white squares: **6**

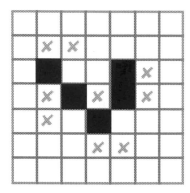

Fig. 2.21   Solution to Retrolife puzzle (b).
Target number of marked white squares: **9**

Fig. 2.22   Solution to Retrolife puzzle (c).
Target number of marked white squares: **13**

Fig. 2.23   Solution to Retrolife puzzle (d).
Target number of marked white squares: **19**

2. The shape is its own predecessor!

3.

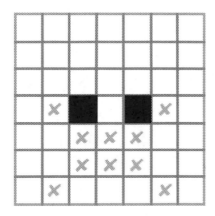

Fig. 2.24   Solution to Retrolife puzzle with a different condition

4.

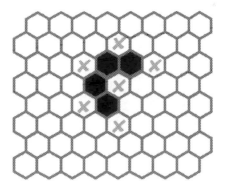

Fig. 2.25   Solution to hexagonal Retrolife puzzle

Bibliography and Further Reading

Adamatzky, A. (Ed.) (2002). *Game of Life Cellular Automata*. Springer-Verlag, London.

Ashbacher, C. (2007). Retrolife generation of the twelve pentominoes, *Journal of Recreational Mathematics*, **36**(1), pp. 35–41.

Berlekamp, E. R., Conway, J. H., & Guy, R. K. (2018). *Winning Ways for Your Mathematical Plays: Volumes 1–4*, CRC Press, USA.

Chow, T., Lucking-Reiley, D., Lucking-Reiley, M. K., et. al. (1996). 1996 Mystery Hunt, *MIT Mystery Hunt.*
https://www.mit.edu/~puzzle/1996/mysteryhunt1996.pdf

Conway, J., & Kochen, S. (2006). The free will theorem, *Foundations of Physics*, **36**(10), p. 1441.

Elran, Y. (2011). Retrolife, in Pegg, E. Jr., Schoen, A. H. & Rodgers, T. (Eds.), *Homage to a Pied Puzzler*. A. K. Peters, Wellesley, MA, pp. 129–136.

Elran, Y. (2012). Retrolife and the pawns neighbors, *The College Mathematics Journal*, **43**(2), pp. 147–151.

Elran, Y. (2020). A generalization of Retrolife, in Plambeck, T., & Rokicki, T. (Eds.), *Barrycades and Septoku: Papers in Honor of Martin Gardner and Tom Rodgers, Spectrum Vol.* 100. AMS/MAA Press, Providence, RI, pp. 131–138.

Gardner, M. (1983). *Wheels, Life and Other Mathematical Amusements*. W. H. Freeman, New York.

Wolfram, S. (2002). *A New Kind of Science.* Wolfram Media, Champaign, IL.

chapter

3

Number Search, Graeco–Latin Square Puzzles and Miracle Sudokus

Introduction

This chapter is all about finding patterns and filling in numbers. Sudoku entered into our lives at the turn of the century, mainly due to Wayne Gould, a computer scientist who wrote a computer program that could quickly create unique puzzles. Little is it known that, apart from being around much longer, it also has origins in a much earlier concept — the Latin square from which stems the idea of Graeco–Latin or Euler squares.

Contrariwise, Word Search, where you have to find given words hidden in a grid amongst random letters, is a puzzle that was invented only in the late sixties of the 20th century, but quickly became a very popular pastime. Figure 3.2 shows a typical Word Search puzzle whose theme is vehicles, and its solution. You may want to try and solve the puzzle on your own before peeking at the answer. There are seven hidden vehicles. To make it more challenging, we haven't given you the words.

	8			3		6		4
			7				8	3
1							9	
		9	5	7		3	1	
	5	8		9	4	2		
	4							1
7	3				8			
8		1		6		4		

5	8	2	9	3	1	6	7	4
6	9	4	7	2	5	1	8	3
1	7	3	8	4	6	5	9	2
4	6	9	5	7	2	3	1	8
2	1	7	6	8	3	4	5	9
3	5	8	1	9	4	2	6	7
9	4	6	2	5	7	8	3	1
7	3	5	4	1	8	9	2	6
8	2	1	3	6	9	7	4	5

Fig. 3.1   A Sudoku puzzle and solution

t	r	a	i	n	i
i	r	i	l	a	s
t	t	r	a	m	t
a	c	p	m	o	a
l	p	l	i	n	x
s	u	a	e	h	i
u	o	n	e	l	s
b	z	e	r	a	c

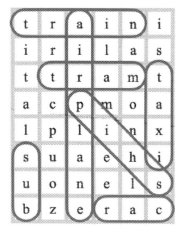

Fig. 3.2   Word Search puzzle and solution

A Graeco–Latin square puzzle is a combination of a Sudoku-style puzzle and Word (or in truth, Number) Search where you are asked to find certain structures, "Graeco–Latin squares", and then fill in their missing numbers. For the purposes of the puzzle we'll define a Graeco–Latin square as a grid of 2-digit numbers where:

- the first digit appears only once in every row and every column,

- the second digit appears only once in every row and every column, and

- each 2-digit number is unique, that is to say no 2-digit number repeats itself in the square.

It is customary that in an $n \times n$ board, the digits go from 1 to n ($n > 1$). Figure 3.3 shows an example of a 3×3 Graeco–Latin square using the digits 1, 2 and 3.

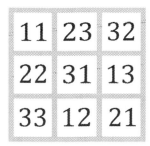

Fig. 3.3   A Graeco–Latin square

The Puzzle

Find the five Graeco–Latin squares hidden in the board in Fig. 3.4 and fill in the missing numbers. Hints: There are no redundant numbers (numbers that are not part of any Graeco-Latin square); two of the Graeco–Latin squares partially overlap.

11				22			43
		13	23				
33				33	41		
25			13	44			
	12		55				
42		35				14	22
		41	22		11		34
14	21			44		21	

Fig. 3.4   Graeco–Latin square puzzle

Where to Start?

To solve this puzzle, you need to do two things: Find and mark all the Graeco–Latin squares and fill in the missing numbers *a la* Sudoku style. What should you do first? There's no real rule, but I prefer to do the search first and find the outlines of at least some of the Graeco–Latin squares. This gives you some clear areas within which you can work. One must remember that, like in Word Search, there *may* be redundant entries — numbers that do not belong to any Graeco–Latin square — so we have to be a bit careful. Empty cells though, cannot hold redundant numbers, so their dispersion gives us a clue as to the boundaries of the Graeco–Latin squares. To keep things simple, in the puzzle above, we have not included any redundant numbers.

Another key to finding the squares comes from the restrictions on the numbers. For example, wherever you find a 5, you know that it is enclosed within at least a 5 × 5 square. Corners also help a lot. A corner square with a 2-digit entry is the corner square of a Graeco–Latin square the size of at least its largest digit, unless of course it is a redundant number.

Solving the Puzzle

Figure 3.5 shows the solution to the puzzle. Graeco–Latin squares are bordered by thick outlines.

11	23	32	31	13	22	14	31	43
22	31	13	23	32	11	23	42	34
33	12	21	12	21	33	41	24	12
25	51	32	13	44	44	32	13	21
31	12	43	24	55	12	24	33	41
42	23	54	35	11	31	43	14	22
53	34	15	41	22	23	11	42	34
14	45	21	52	33	44	32	21	13

Fig. 3.5   Solution to Graeco–Latin square puzzle

Note that two of the squares overlap and that there are no redundant numbers in this puzzle.

The History of the Puzzle and Related Topics

Ah! The satisfaction of solving a puzzle — that's what drives a lot of us to spend time on things that seem practically useless, be it searching a grid for a list of words, or filling in Sudokus. Since I enjoy both, I decided to combine them into one puzzle — the Graeco–Latin square puzzle. I introduced this puzzle at a conference in 2008, so it doesn't quite have a lot of mileage, yet. But Sudoku and Word Search do, so it's worthwhile taking a look at their history.

Word Search is attributed to two puzzlers, who seem to have invented it independently: the late Spanish crossword and puzzle master, Pedro Ocón de Oro; and Norman E. Gibat from Norman, Oklahoma (co-incidentally the same city where Martin Gardner lived in his older days). Ocón de Oro was a prolific puzzler, inventing over 125 different kinds of puzzles and publishing them regularly in Spanish and Latin American books and newspapers. His Word Search, which seems to have been invented before Gibat's was called *Sopa de Letras* (Soup of Letters). Ocón de Oro was a linguist so most of his puzzles were language-based. Having won the crossword puzzle at the age of 16, he was certainly qualified. There seems to be an intriguing relationship between linguistics, music and math, and often people who like math are also competent musicians and linguists. Norman E. Gibat was actually a publisher. He first published Word Search in his classified-listings newspaper, the *Selenby Digest*. Obviously, he liked word play — the name of his newspaper is a play on the words "sell and buy".

Since its invention, there have been many generalizations on Word Search — too many to count — but you can find some in the following section and the puzzles at the end. The reason there have been so many different branches of Word Search is probably because the initial puzzle *seems* to be not very "intelligent". After all, it's just "searching". However, I think that it might be just that which appeals to so many people who sometimes want an easier challenge, or just want to sharpen their observation skills.

Which brings us to Sudoku. The word Sudoku in Japanese is an acronym of *Suuji Wa Dokushin ni Kagiru* (which means "the numbers must be single"), alluding to the basic specifications of the puzzle — that each of the digits 1 to 9 appear exactly once in every row, column and 3×3 major sub-square of the grid. Funnily enough though, Sudoku didn't originate in Japan! The first Sudoku puzzle, at least its present-day form, was published by Howard Garns, an architect, in the May 1979 issue of *Dell Pencil Puzzles & Word Games*. In the 1980's, Nicoli, a Japanese puzzle company discovered Dell's riddle and improved on it. Kaji Maki, the president of Nicoli, then renamed it SuDoKu. The puzzle became a great success in Japan, and was then reintroduced to the West in 2004 by Wayne Gould, a retired Hong Kong judge. Gould wrote a computer program that generated Sudoku puzzles and managed to convince *The Times* to publish them. He did this free of charge, refusing to accept any royalties.

Sudoku's predecessor is the Latin square, that has been around for thousands of years and was commonplace in ancient China and in Europe during the Middle Ages. It was used mostly as amulets with words or symbols instead of numbers. A Latin square is a grid of numbers where each number appears once in each row and once in each column, so it is really a Sudoku without the extra "sub-square" condition.

Leonhard Euler, the prominent 18th century Swiss physicist and mathematician was the first to study Latin squares, from a mathematical perspective. Born in 1707, he lived in Russia and Germany, and published many scientific papers (886 in his lifetime). It is said that he could solve any problem, and indeed there are between 60 to 80 mathematical concepts that are named after him. Euler coined the term Latin square, since he filled his squares not with numbers, but with the ordered letters in the Latin alphabet. Euler's credentials were so high that French mathematician and physicist Pierre-Simon Laplace used to preach to his students, "Read Euler, read Euler, he is the master of us all."

Since Euler's time, Latin squares have become popular not only with sorcerers and talisman artists, but with mathematicians, scientists and the general public. It became customary to fill the Latin squares with numbers from 1 to n, where n is the size of the square (which we call of order n). Mathematicians turned their interest

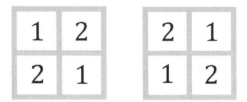

Fig. 3.6   The two 2 × 2 Latin squares

to trying to find all possible Latin squares of a given size. It was straightforward enough to define the unique 1 × 1 Latin square (just the number 1), and find the two 2 × 2 Latin squares (Fig. 3.6), the twelve 3 × 3 squares, and the 576 squares of order 4. However, as the size got bigger, the feat became much more difficult, since the number of Latin squares grew exponentially, and it was only in the 20th century that mathematicians, or rather computer scientists, were able to pinpoint the number for larger squares, using brute force computations to enumerate all possible configurations. To give you an idea of the sheer magnitude of these numbers, there are: 5,524,751,496,156,892,842,531,225,600 Latin squares of *order* 9, of which 6,670,903,752,021,072,936,960 are Sudokus.

How do we know these numbers? Mathematicians have not yet found a formula to compute the exact number of possible Latin squares (or Sudokus larger than 9 × 9) of a given size, though they have succeeded in finding upper and lower bounds for the numbers of Sudokus and Latin squares of a given order, using mathematical reasoning. What's left to do then, is to use computers that can in principle find the exact numbers by exhausting all possibilities. However, computers have only been able to calculate the number of Latin squares up to order 11 so far (as of December 2020).

How do *I* know these numbers? One of the best places to get information like this is Neil Sloane's *On-Line Encyclopedia of Integer Sequences*®. I really recommend taking a look at this phenomenal database of sequences of numbers, where you can learn about almost all conceivable "counts" of mathematical objects there is to find.

Back to Euler. Euler didn't stop with regular Latin squares. He went on to discover what are known as Graeco–Latin squares, although Euler didn't use that name (in fact, he didn't name his squares at all!). The two-symbol entries in the squares, as their name suggests, were Latin letters for the first symbol, and Greek for the second (note that this is the opposite order of what the name Graeco–Latin suggests). Figure 3.7 shows a 5 × 5 Graeco–Latin square.

Aα	Bδ	Cβ	Dε	Eγ
Bβ	Cε	Dγ	Eα	Aδ
Cγ	Dα	Eδ	Aβ	Bε
Dδ	Eβ	Aε	Bγ	Cα
Eε	Aγ	Bα	Cδ	Dβ

Fig. 3.7   A 5 × 5 Graeco–Latin square

Graeco–Latin squares come in many sizes, but not all. Euler knew of course that you cannot construct a 2 × 2 Graeco–Latin square. He managed to find 3 × 3, 4 × 4 and 5 × 5 Graeco–Latin squares, but couldn't find any 6 × 6 squares, leading him to hypothesize that Graeco–Latin squares exist only if their order is either an odd number or an even number divisible by 4 (i.e. squares of sizes 6 × 6, 10 × 10, 14 × 14, and so on, don't exist). This hypothesis was thought to be true for almost 200 years. It was only in 1959 that three mathematicians, Raj Chandra Bose, Sharadchandra Shankar

Fig. 3.8   Portrait of Leonhard Euler by Jakob Emanuel Handmann (1753)

Shrikhande, and Ernest Tilden Parker, proved Euler right about the non-existence of 6×6 Graeco–Latin squares, but totally wrong about *all* the rest! It has been *proven* that Graeco–Latin squares of all orders *except* 2 and 6 exist, which teaches us a lesson: don't try to generalize your theorem too early! As a tribute to Leonhard Euler, Graeco–Latin squares also became known as Euler squares.

Both Latin squares and Euler squares serve many practical purposes, in particular in experiment design. For example, suppose we want to test which of five species of flowers will grow best in a field. We want the conclusions that we derive from our experiment to be as accurate as possible. We could just divide the field into five rows, plant a different species of flower in each row, and see what happens, but we risk coming to the wrong conclusion if the quality of the soil is not distributed isotropically among different rows. However, if we divide the field into 5×5 patches, and plant the flowers according to a 5×5 Latin square design, we ensure a fair experiment, since every species is tried in all the different rows in the field.

We can even add another parameter to the same experiment. Suppose we have five different brands of fertilizer and would like to find out which fertilizer yields the best results, we can use a Graeco–Latin square to find the answer. This way we can be sure that each of the five fertilizers was paired with each of the five species of flowers and in each of the five rows in the field.

Generalizations of the Puzzle

There is so much math you can learn from Latin squares and its variations. We have already seen its uses in experiment design. One of the reasons is that the concept is quite general, leaving room for many avenues of math exploration. Here are some of the things you can do with Latin squares.

First, you can create new ones! One way to do this is by switching any two rows or columns in a given Latin square. Or, you can exchange numbers, for example, change all 3's to 1's, all 1's to 2's and all 2's to 3's. The set of Latin squares that you get from these operations are called *isotopic Latin squares.* Interestingly, all twelve 3×3 Latin squares form a single isotopic set.

Another cool thing to do is to make a magic square from a Latin square. In fact, it has been suggested that Euler's work on Latin squares actually started because of his interest in magic squares, which were extremely popular during the Middle Ages. A magic square is a square where the numbers in each row, column and main diagonals sum up to the same total. One famous magic square appears in Melancholia, a wooden engraving made by German artist Albrecht Dürer in 1514 (Fig. 3.9) that now resides in the Metropolitan Museum of Art. The magic square appears in the upper right corner of the engraving. Figure 3.10 shows the magic square enlarged. Notice that every row, column and main diagonals add up to 34, the so-called "magic number".

Dürer's magic square is a typical magic square, but not a Latin square, since each number appears only once. To turn a Latin square into a magic square, with each number appearing exactly n times, where n is the order of the square, what you need to do is make sure

Fig. 3.9   Albrecht Dürer's engraving, Melancholia I

Fig. 3.10   The magic square in Melancholia I

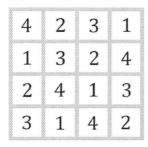

Fig. 3.11   A magic Latin square of order 4

that each number on the diagonal appears only once. Figure 3.11 shows a magic Latin square of order 4.

There *is* a way to turn a Latin square, or more precisely a Graeco–Latin square into a magic square where *no* numbers repeat. In fact, this is exactly what Euler started off trying to do when he began his work. What you need to do is to take a numeric Graeco–Latin square, like the 5 × 5 square on the left-hand side of Fig. 3.12, and replace each entry in ascending numeric order by the numbers 1, 2, 3, and so on, until every number is replaced. In the square on the left-hand side of Fig. 3.12, the lowest number is 11 and it is replaced by 1; the next number up, 12, is replaced by 2; and so on until the largest number 55 is replaced by 25. You can see the new square in the middle of Fig. 3.12. This is still not a complete magic square. All the rows and columns *do* add up to the same sum, the

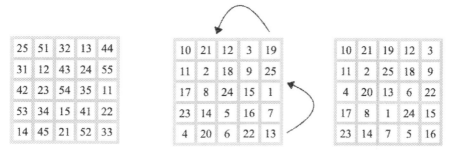

Fig. 3.12   From Graeco–Latin square (left) to non-diagonal magic square (center) to full magic square (right)

magic number 65, but one of the main diagonals does not, so you need to rearrange the rows and columns to get the right sums on both the main diagonals as well. This will not change the sums in the rows and columns, so the result will be a 5 × 5 magic square (on the right-hand side of Fig. 3.12). There are many ways you can rearrange the rows and columns to get the diagonals right. One easy way that works for an odd square, is to rearrange the rows and columns so that the number in the center cell is the magic number divided by the order of the square; in our example, that would be $65 \div 5 = 13$.

To complete the picture, perhaps it should be mentioned that a Graeco–Latin square can be decomposed into two *orthogonal Latin squares*: one of them with all the first digits of the numbers in the Graeco–Latin square, and the other with the second digits. Figure 3.13 shows a Graeco–Latin square of order 3, decomposed into its two, constituent, orthogonal Latin squares.

Fig. 3.13   An order 3, Graeco–Latin square decomposed into orthogonal Latin squares

Conversely, two Latin squares are orthogonal if, when superimposed on each other, all the (new) entries are distinct. We can even construct sets of *mutually orthogonal Latin squares,* known as MOLS, where every pair of Latin squares is orthogonal. There is only one such set of order 5 squares consisting of four Latin squares. The number of MOLS of order 10 and above is still not known. The math of "counting things" is an active field of research in combinatorics.

Isotopic, orthogonal and Graeco–Latin are all special aspects of Latin squares, and they're all very interesting, but the queen of Latin squares is undoubtedly Sudoku. The number of variations in Sudoku is so humongous, that it is impossible to list everything here. Wikipedia has a glossary of Sudoku, where many variants can be found. One of the more popular variants is to change the grid topology from a 9×9 grid divided into 9 square regions, to any $n \times n$ grid divided into n regions, not necessarily square. Figure 3.14 shows a giant Sudoku (appropriately known as Sudoku the Giant),

J	4	N								C			B	2	M	P			E		H			O
H	D		O		6						8			1	A	B	G	C	E	5	L			F
	8		I		A	K	O	3	B	M	L	F	5	1		H	7			C			6	J
B			A					G	L	N	J				H	6	8			D	M	1	2	7
	L	1	5		M		4	2	N			P			D	J				6	9	B	8	A
F	H				N	O	4	5			D				M	J		I		6		9	C	8
5					M		6	F		K	9	A	C							1		L		
	1					I	2				J	K		7	A	B			N		H	O		
6	A				E	G	9			C			L	O	2	5	7	1	8	F		J	K	M
I	J				K	D	L					1			E	G		3	H				B	5
M	5	3	L	7	N	A	C	I			F	B		G	K	E				O	2	J	H	
	F					B	G			O			1	9	E		7			L	5	K	D	6
K							1			5	O	H			6				9		N			
D	G				J	5	H	3		K	P				B				N	1	C	E	8	
1			C		B	7	F	6	K	D	2		M		N		4			J			5	9
L	I			5			A	E		B			1	7	F		N	J			C			D
8	6	A	H					C	O					I							F	5	7	
3	C	B	1			L		F	9		A	4				7	8	2	N				6	
	E	G					7			1	5	C			L		2				H			K
	F				O							H	J			4	C		D	3	E	I	1	L
	N	6	F	H		M	E	K	3		9	P									G	O	2	
G	O	5	3	C	P		E	8		F			6						4	B	J	7		I
9	I	D	8	L	B			6				G			4	H	5	J		C	A		F	1
	J				1	G		F	7	5	9	N	L			2	A			6				C
	B				C		9					A				G		8				K	D	E

Fig. 3.14   Sudoku the Giant

J	4	N	7	6	1	9	D	5	8	A	C	G	B	2	M	P	L	F	3	E	K	H	I	O
H	D	M	O	2	6	P	I	7	J	K	8	4	9	1	A	B	G	C	E	5	L	3	N	F
9	8	G	I	F	A	K	O	3	B	M	D	L	F	5	1	N	H	7	2	C	P	4	6	J
B	3	K	A	P	F	C	G	L	E	N	J	O	H	6	8	4	9	5	1	D	M	1	2	7
C	L	1	5	F	M	H	4	2	N	3	E	P	I	7	O	D	J	K	6	9	B	8	A	G
F	H	L	N	O	4	5	7	A	1	D	2	E	G	M	J	K	I	B	P	6	3	9	C	8
5	7	8	B	M	E	6	F	G	P	H	3	I	J	K	9	A	C	4	O	2	1	D	L	N
4	1	9	C	3	I	2	8	J	K	6	7	5	A	B	L	M	F	D	N	G	H	O	E	P
6	A	D	E	G	9	3	B	C	H	L	4	N	O	P	2	5	7	1	8	F	I	J	K	M
I	J	P	2	K	D	L	M	N	O	8	1	9	C	F	E	G	6	3	H	7	4	A	B	5
M	5	3	L	7	N	A	C	I	9	4	F	B	6	G	D	8	K	E	1	P	O	2	J	H
N	F	H	J	A	B	G	P	O	4	1	9	C	8	E	I	7	2	L	5	K	D	6	M	3
K	2	4	P	I	8	E	1	M	L	5	O	H	D	J	6	C	3	9	G	N	7	B	F	A
D	G	O	6	9	2	J	5	H	3	7	I	K	P	L	B	F	A	N	M	1	C	E	8	4
1	E	C	8	B	7	F	6	K	D	2	A	M	3	N	H	O	4	P	J	I	G	L	5	9
L	I	2	M	5	3	4	A	E	6	B	G	1	7	8	F	H	N	J	K	O	9	C	P	D
8	6	A	H	4	K	M	9	D	C	O	N	2	E	3	P	I	1	G	L	J	F	5	7	B
3	C	B	1	J	H	I	L	P	F	9	5	D	K	A	4	E	O	M	7	8	2	N	G	6
O	P	E	G	D	J	7	N	1	5	C	6	F	L	I	3	2	B	8	9	H	A	M	4	K
7	K	F	9	N	O	8	2	B	G	P	H	J	M	4	C	6	5	A	D	3	E	I	1	L
A	N	6	F	H	5	D	J	4	M	E	K	3	1	C	7	9	P	I	B	L	8	G	O	2
G	O	5	3	C	P	N	E	8	A	F	L	6	2	D	K	1	M	H	4	B	J	7	9	I
P	9	I	D	8	L	B	K	6	2	G	M	7	4	H	5	J	E	O	C	A	N	F	3	1
E	M	J	K	1	G	O	3	F	7	I	B	8	5	9	N	L	D	2	A	4	6	P	H	C
2	B	7	4	L	C	1	H	9	I	J	P	A	N	O	G	3	8	6	F	M	5	K	D	E

Fig. 3.15   Solution to Sudoku the Giant

and Fig. 3.15 its solution (you might like to solve the puzzle yourself before looking at the solution). Note the letters in Sudoku the Giant. This is because each row, column and region holds 25 symbols — the ten digits 0–9, and the letters A through O. Other Sudoku variations of this kind can be found in the Challenge Yourself! section at the end of this chapter.

Another favorite variation is to add the restriction that the numbers on the main diagonals are also all different. This turns the Sudoku/Latin square into a magic square. My favorite, recently highlighted on Mark Goodliffe and Simon Anthony's "Cracking the Cryptic" YouTube channel, is Mitchell Lee's Miracle Sudoku. You'll find one of these in the Challenge Yourself! section. Played on a regular Sudoku grid, the remarkable thing about this Sudoku is that the initial board has only two entries! If the rules were the regular Sudoku, there would be a huge number of solutions to this puzzle. However, Lee added some extra rules that make it unique: a number cannot appear in any cell that is a (chess) knight's move away from its own position. This means that the prohibited cells are at a distance of two squares horizontally and one square vertically, or two squares vertically and one square horizontally. Similarly, a number cannot be at a king's move away from its own position, that is, a number cannot repeat in any of the adjacent eight cells (horizontal, vertical and diagonal). Additionally, two consecutive numbers, like 1 and 2, or 2 and 3, cannot be in adjacent horizontal or vertical cells (but they can be adjacent diagonally). Using logic alone, these Sudokus can actually be solved! You can see Lee's miracle Sudoku puzzle and solution in Fig. 3.16.

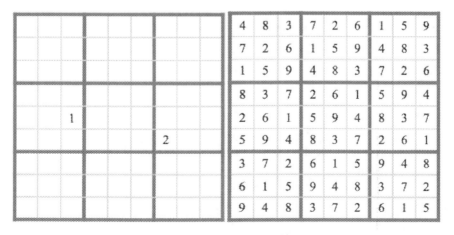

Fig. 3.16   Miracle Sudoku and solution

Math and computer science help us create and solve Sudokus. There are quite a few ways to construct Sudokus. The most common way can be briefly described as follows. First create a fully solved Sudoku grid like this:

1. Randomly place the nine numbers 1–9 in an empty 9 × 9 grid.

2. Write in each of the other 72 cells, all the numbers from 1–9.

3. From each of these cells, erase those numbers that are in the same line, column or 3 × 3 region as the corresponding numbers in the single number cells (those filled in at the beginning).

4. Randomly select a cell in turn and remove all the numbers except one. Then erase that number from the cells that are in the same row, column or sub-square.

5. Repeat steps 1 to 4 until you have a fully-solved Sudoku board, or if you reach a dead end, backtrack (return to the last point where you made a choice — this would be either the number you chose to remain in the randomly-chosen cell in step 4, or if you've exhausted all possibilities, the random cell itself — and make a different choice).

Once there is a full board, more entries need to be deleted to create the puzzle. This can be done randomly, checking all the time that the solution remains unique. There is of course, also the option of transforming a given 9 × 9 Latin square into a Sudoku by exchanging rows and columns until the extra restriction of 3 × 3 regions is observed. Note, however, that this is not always possible, and once you have the Sudoku arrangement you still have to remove entries to get the puzzle.

One strategy for *solving* Sudokus is to detect spaces in which a certain number must appear. This can be done as follows. Pick a certain number that appears on the board, draw a straight line through all the rows and columns in which that number appears, and check if you are left with just one empty space in one of the small 3 × 3 squares. If so, this is where the number has to go. Do this for all the numbers from 1 to 9, iteratively.

Another common strategy uses logical reasoning. For example, we might know what numbers should be in some of the squares, but we don't specifically know which number to put in each square. Still, this may be enough information to solve for a certain entry. In Fig. 3.17 we know that the numbers 3, 5 and 6 should be written in the three middle squares of the top row, to complete the middle 3 × 3 square, though we don't know the order. We can infer that the empty square on the right-hand side of the top row has to be 4, the only other number left to fill the top row. There are, of course, many other strategies, and sometimes, backtracking is needed, but this is what makes Sudoku such a great puzzle.

9	1	7				2	8	
			7	1	9			5
			2	8	4		7	9

Fig. 3.17   Use logical reasoning to infer the missing number in the top-right corner

Counting Sudokus seems to be a favorite pastime in combinatorics, as bears witness the many Sudoku-related entries in Neil Sloane's *On-line Encyclopedia of Integer Sequences*. Many, as yet unknown, problems wait to be solved. For example, we know there are 288 Sudokus of order 4, and 6,670,903,752,021,072,936,960 Sudokus of order 9, but that's it! We still haven't been able to work out how many 16 × 16 Sudokus there are. We know that 17 is the minimum number of clues (entries on a puzzle board) needed to have a unique solution for a regular 9 × 9 Sudoku, but we don't know the minimum number for larger boards, though we do know that the number for a 16×16 Sudoku is less than or equal to 55. And the list of open problems goes on.

Word Search puzzles also have many variations. One is Word Snake, where the words "snake" through the grid. Unlike Word Search, the words hidden in Word Snake sometimes turn a few times, at an angle of 90°. Another popular variation is when all the hidden words have been recovered, the unused letters spell out a secret message. Number searches are also quite popular. They're exactly the same as Word Search, except that you have to find numbers instead of words.

We finish off this chapter with one of the latest additions to the genre — Hidato. Hidato is not strictly a number search puzzle, but it does resemble a Graeco–Latin square puzzle in a certain sense. Hidato was invented a few years ago by Israeli computer scientist Gyora Benedek, who was inspired while scuba diving in Eilat. Hidato became quite popular, and is now distributed in puzzle columns in major newspapers, puzzle books (like *Hidato Beehive*) and dedicated websites and mobile apps (see Bibliography section). Figure 3.18 shows a Hidato puzzle and its solution. This puzzle is shaped as a

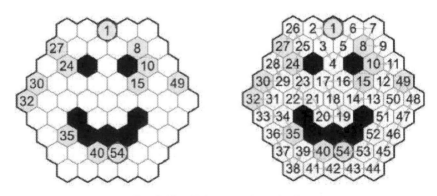

Fig. 3.18   Hidato puzzle and solution

"hexagon of hexagons" to be filled in with all the whole numbers between 1 and 54, under the restriction that consecutive numbers are filled in adjacent hexagons.

Recap

This chapter gives us a glimpse into the world of word and number search puzzles. Latin and magic squares that sparked the interest of famed mathematician Leonhard Euler, led the way to his construction of Graeco–Latin squares, used even nowadays for designing experiments. They also are the precursors of Sudoku and its many variations, including the very recent Miracle Sudokus. There are many open questions regarding Sudoku, mainly in the field of combinatorics, yet to be solved. When Graeco–Latin squares are combined with number search puzzles, the Graeco–Latin square puzzle arises. Along the way, we saw quite a few variations on these topics: isotopic, orthogonal and mutually orthogonal Latin squares, magic, giant and monster Sudoku, Word Snake and Hidato. We saw how to turn Graeco–Latin squares into magic squares, and learnt that you can create new Latin squares by switching any two rows or columns, or by exchanging the numbers in a given Latin square. Finally, we also saw some algorithms used to create and solve Sudokus.

Here are short, informal definitions of some of the terms in this chapter:

- *Combinatorics* — an area in math that studies ways of counting things.

- *Graeco–Latin square* — a square made by superimposing two orthogonal Latin squares. Also known as a Euler square.

- *Isotopic Latin square* — a Latin square is isotopic to another Latin square if the original square can be restored by switching rows and/or columns and/or substituting every fixed number or symbol with a different one from the same set.

- *Latin square* — a grid of numbers or symbols where each number (symbol) appears once in each row and once in each column.

- *Magic square* — a square where the numbers in each row, column and main diagonals sum up to the same total.

- *Mutually orthogonal Latin squares (MOLS)* — a set of Latin squares in which every pair is orthogonal.

- *Order of a Latin square* — the number of rows or columns in a Latin square.

- *Orthogonal Latin squares* — two Latin squares of the same order are orthogonal if, when superimposed on each other, all the (new) entries are distinct.

- *Sudoku* — a puzzle in which a partially filled-in $n \times n$ (n usually 9) Latin square of numbers 1 through n, has to be completed. In addition to the usual Latin square constraints, in Sudoku, there are also sub-regions (typically 3×3 squares) in which each of the numbers 1 through n appears exactly once.

- *Word Search* — a puzzle where you have to find some given words hidden in a grid amongst random letters.

Challenge Yourself!

1. Solve the following prime number search: find these 18 prime numbers hidden in the grid:

1 0 0 9
2 0 6 9
1 9 7 9
4 0 7 3
4 5 8 3
8 3 5 3
9 7 9 1
5 4 0 7
5 6 6 9
4 0 9 3
9 3 3 7
5 1 6 7
6 6 0 7
2 0 8 7
1 5 1 1
1 1 5 1
9 0 0 1
6 5 2 1

1	2	5	6	6	9
5	0	4	1	0	3
1	6	0	5	6	3
1	9	7	9	4	7
8	3	3	6	0	8
5	1	5	6	9	0
2	0	6	3	3	2
5	0	4	5	8	3

Fig. 3.19   Prime number search puzzle

2. Solve the following easy and hard Sudoku puzzles:

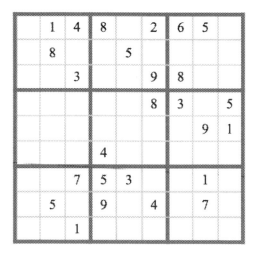

Fig. 3.20 Easy and hard Sudoku puzzles

3. Solve the following Miracle Sudoku: a number cannot appear in any cell that is a chess knight's move or a king's move away from its own position. Consecutive numbers cannot be in adjacent horizontal or vertical cells.

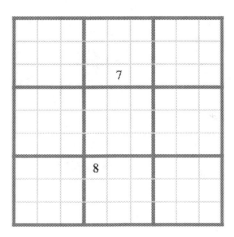

Fig. 3.21   Miracle Sudoku puzzle

4. Fill in the following Latin square such that each number appears exactly once on the main diagonals:

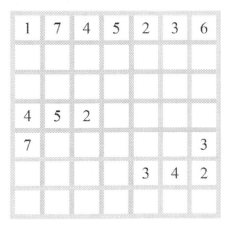

Fig. 3.22   Latin square puzzle

5. Solve the following Hidato puzzles:

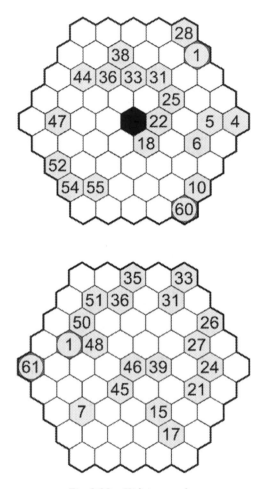

Fig. 3.23   Hidato puzzles

6. Find the six Graeco–Latin squares hidden in the following Graeco–Latin square puzzle.

31							54
	11		21			43	
	32			13	25		
							41
		12					
11							
					34		
	31	24			13		
	43			31			23

Fig. 3.24   Graeco–Latin square puzzle

Solutions

1. Note that some of the bubbles can be read in both directions, i.e. 1511 and 1151.

1 0 0 9

2 0 6 9

1 9 7 9

4 0 7 3

4 5 8 3

8 3 5 3

9 7 9 1

5 4 0 7

5 6 6 9

4 0 9 3

9 3 3 7

5 1 6 7

6 6 0 7

2 0 8 7

1 5 1 1

1 1 5 1

9 0 0 1

6 5 2 1

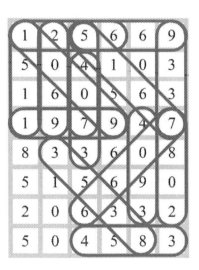

Fig. 3.25   Solution to the prime number search puzzle

2.

7	9	6	5	2	4	1	8	3
1	3	8	9	7	6	2	5	4
2	5	4	8	1	3	7	6	9
3	8	1	7	6	9	5	4	2
5	4	2	1	3	8	6	9	7
6	7	9	4	5	2	3	1	8
4	2	5	3	8	1	9	7	6
9	6	7	2	4	5	8	3	1
8	1	3	6	9	7	4	2	5

9	1	4	8	7	2	6	5	3
7	8	2	6	5	3	1	4	9
5	6	3	1	4	9	8	2	7
1	4	9	7	2	8	3	6	5
2	7	8	3	6	5	4	9	1
6	3	5	4	9	1	7	8	2
8	2	7	5	3	6	9	1	4
3	5	6	9	1	4	2	7	8
4	9	1	2	8	7	5	3	6

Fig. 3.26   Solutions to the easy and hard Sudoku puzzles

3.

6	3	9	2	8	5	7	4	1
1	7	4	6	3	9	2	8	5
5	2	8	1	7	4	6	3	9
9	6	3	5	2	8	1	7	4
4	1	7	9	6	3	5	2	8
8	5	2	4	1	7	9	6	3
3	9	6	8	5	2	4	1	7
7	4	1	3	9	6	8	5	2
2	8	5	7	4	1	3	9	6

Fig. 3.27   Solution to the Miracle Sudoku puzzle

4.

1	7	4	5	2	3	6
3	2	1	6	5	7	4
6	3	7	2	4	5	1
4	5	2	3	1	6	7
7	4	5	1	6	2	3
5	1	6	7	3	4	2
2	6	3	4	7	1	5

Fig. 3.28   Solution to the Latin square puzzle

5.

Fig. 3.29   Solutions to the Hidato puzzles

6.

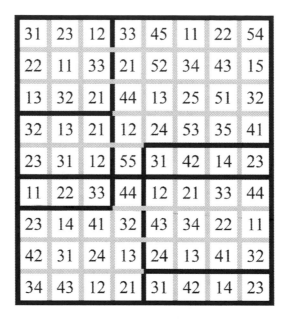

Fig. 3.30   Solution to the Graeco–Latin square puzzle

Bibliography and Further Reading

Benedek, G. (2020). Hidato. https://www.hidato.com

Berend, D. (2018). On the number of Sudoku squares, *Discrete Mathematics*, **341**(11), pp. 3241–3248.

Bose, R., Shrikhande, S., & Parker, E. (1960). Further results on the construction of mutually orthogonal latin squares and the falsity of Euler's conjecture, *Canadian Journal of Mathematics*, **12**, pp. 189–203.

Fasino, D. (2020). Sudoku_lite, *MATLAB Central File Exchange.* Retrieved December 26, 2020, from https://www.mathworks.com/matlabcentral/fileexchange/9462-sudoku_lite (software used to generate Sudokus).

Glossary of Sudoku. Wikipedia. Retrieved December 26, 2020, from
https://en.wikipedia.org/wiki/Glossary_of_Sudoku

Goodlife, M., & Anthony S. (2020). Miracle Sudoku by Lee Mitchell, *Cracking the Cryptic*, Retrieved January 7, 2021, from
https://youtu.be/yKf9aUIxdb4

McNair, R. (2005). Number of (completed) Sudokus (or Sudokus) of size $n^2 \times n^2$, *The On-line Encyclopedia of Integer Sequences*. Retrieved December 26, 2020, from https://oeis.org/A107739

Nacin, D. (2019). Puzzles, parity maps, and plenty of solutions, in Beineke, J., & Rosenhouse, J. (Eds.), *The Mathematics of Various Entertaining Subjects, Volume 3, The Magic of Mathematics.* Princeton University Press, Princeton, NJ, pp. 222–246.

Pegg, E. Jr., & Weisstein, E. W. (n.d.). Sudoku, from *MathWorld*—A Wolfram Web Resource. Retrieved December 26, 2020, from
https://mathworld.wolfram.com/Sudoku.html

Wilson, R., & Watkins, J. J. (Eds.) (2015). *Combinatorics: Ancient and Modern*. Oxford University Press, Oxford.

Operator Puzzles

Introduction

Many number puzzles are designed so that you have to find missing numbers. Sudoku and Kakuro are great examples. Another is cryptarithms. Generally speaking, a cryptarithm is an encrypted arithmetic expression, where each digit is replaced by a letter. Using logic, one can "crack" the code and recover the expression. The solution to the cryptarithm

$$BAD + FOOD = MAKES$$

is

$$807 + 9557 = 10364 .$$

Operator puzzles are similar to cryptarithms except that what's encrypted in the mathematical expression is the operator — the symbol that signifies what's to be done to the numbers to get the correct answer. In the easiest operator puzzles the encrypted operators are usually the symbols for addition, subtraction, multiplication and division: $+, -, \times, \div$, but more advanced puzzles include less obvious operators, like the square root, or the factorial (both incidentally operate on only one number), and even made-up operations, for example, the sum of the digits of the number. Note that the word *operator* is used for the symbol, while the actual action taken is called the *operation*. Hence, the factorial *operator*, denoted by an exclamation mark, performs the following *operation* on a given number: it multiplies all the whole numbers up to the given number, for example,

$$6! = 1 \times 2 \times 3 \times 4 \times 5 \times 6 = 720.$$

In many puzzles, we immediately have a problem, since there can be more than one correct operator. An example is the following puzzle, where the question mark is to be replaced with an operator:

$$2 ? 2 = 4.$$

The answer can be

$$2 + 2 = 4,$$

but it can also be

$$2 \times 2 = 4.$$

Therefore, a few expressions with the same missing operator are usually given, so that the solution is unique. It is obvious that in the following series of expressions the missing operator is indeed "+":

$$2 ? 2 = 4$$
$$6 ? 10 = 16$$
$$8 ? 33 = 41$$

So ? = "+" and the answers are

$$2 + 2 = 4$$
$$6 + 10 = 16$$
$$8 + 33 = 41.$$

The Puzzle

Find the missing operators in the following puzzles. The missing operator is always denoted by a question mark (?).

1. $$13 ? 4 = 52$$
    $$11 ? 34 = 374$$
    $$0 ? 4 = 0$$

2. $$? 4 = 2$$
    $$? 2209 = 47$$
    $$? 0 = 0$$

3. $$1 ? 4 = 41$$
    $$12 ? 3 = 312$$
    $$7 ? 7 = 77$$

Where to Start?

The first two puzzles are operators that you should know, and they appear in the list above. You might like to refresh your knowledge about how the one-number operators work. Operators that operate on one number, like "taking the square root", are called *unary operators*. Operators that operate on two numbers are called *binary operators*. There are even *ternary operators* that operate on three numbers, but these are not in the puzzles above. You might need to work a bit harder on the last puzzle — it is a "made-up" binary operator.

Solving the Puzzle

1. The first puzzle is easy — the missing operation is "multiplication", so the missing operator is "×". In many cases, a zero in the puzzle is a good place to start because in many operations, zeros exhibit

tell-tale behavior, for example, a zero multiplied by any number returns zero.

$$13 \times 4 = 52$$
$$11 \times 34 = 374$$
$$0 \times 4 = 0$$

2. In this puzzle, a unary operator is missing. We're looking for an operator that rapidly reduces long numbers to short numbers. The square root does the trick!

$$\sqrt{4} = 2$$
$$\sqrt{2209} = 47$$
$$\sqrt{0} = 0$$

3. Made-up operators are much harder to guess. Did you spot the fact that the digits on the left-hand side of the equations, are exactly the same as those on the right-hand side? The missing operator here is: "concatenate the second number with the first".

$$1 \text{ concatenated with } 4 = 41$$
$$12 \text{ concatenated with } 3 = 312$$
$$7 \text{ concatenated with } 7 = 77$$

The History of the Puzzle and Related Topics

I came up with operator puzzles when I was looking for a math puzzle for kids — after they had tried their hands at breaking secret codes and after they had learned about cryptarithms. I don't know why, but people, especially kids, are really intrigued by ciphers, codes and encryptions of all sorts. I suppose it's that inner hunger for knowledge, especially that which is concealed. So after we learned about encrypting and deciphering texts (secret codes) and encrypting numbers (cryptarithms), the only thing left to encrypt were the operators!

Operators have not been around a long time. The first time a symbol was used for an arithmetic operation is attributed to Nicolas d'Oresme, a French bishop, mathematician, philosopher and much more, who used the plus sign as an abbreviation of the Latin word, *et*, to signal addition in his book *Algorismus Proportionum* published sometime between 1356 and 1361 (although it could be a later book copier who did this). The minus sign followed, along with the plus sign again (in print), in *Behende und hüpsche Rechenung auff allen Kauffmanschafft* (*Mercantile Arithmetic* in English), written by the German mathematician, Johannes Widmann, in 1489. Soon after, the symbols for square root, parentheses and the equal sign followed. The British mathematician, William Outread, introduced the multiplication symbol in the early 17th century. It might seem funny that the square root symbol preceded multiplication, but this is probably due to the fact that quite often people didn't use a symbol at all for multiplication. They just wrote the factors one next to the other or used a capital M to designate multiplication (and D for division, R for square root). The other symbols quickly followed, most of them appearing together with new math, for example, the derivative notation, developed in the 17th century by the German mathematician, Gottfried Leibniz, who invented it for use with his newly-discovered calculus. We shouldn't really use the word "his", since there was and still is a huge controversy over who invented calculus — Leibniz, or British physicist, Isaac Newton. Both claimed to have invented it, though Leibniz published his work first, albeit amid claims that he plagiarized Newton's unpublished work. Echoes of the controversy still remain, as some physicists prefer to use Newton's dot notation for derivatives instead of Leibniz's notation!

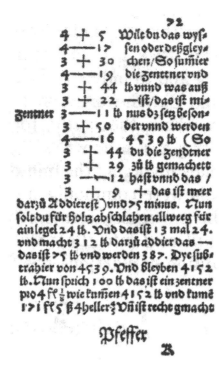

Fig. 4.1 Excerpt from Johannes Widmann's *Behende und hüpsche Rechenung auff allen Kauffmanschafft*, showing the printed plus and minus symbols

Notations for mathematical operations are invented all the time. Sometimes, it's just a matter of luck (or marketing) as to whether someone's invented notation will "stick". One of my favorite recent operator notations is Donald Knuth's up-arrow notation, invented to deal with very, very large numbers. He uses one up-arrow to represent the power operation — exponentiation. This is how he would write 5 to the power of 3:

$$5 \uparrow 3$$

which means

$$5^3 \text{ or } 5 \times 5 \times 5$$

and equals 125. The left-hand operand is the number you operate on and the right-hand operand is the number of times you perform the operation. So, in our example, the operation is repeated multiplications of the number 5, three times. Now comes the interesting part: 5 ↑ ↑ 3 means just the same, except that the operation that is performed three times on the number 5 is the power (↑) meaning five to the power of five to the power of five:

$$5 \uparrow \uparrow 3 = 5^{5^5} = 5^{3125}$$

which is equal to the "monster" number:

19110125979454775203564045597039645991980810489
90094337139512789246520530242615803012059386519
73985026558644015579446223535921278867380697228
84101469159866020879618967571957018392816603380
47611225975533626101001482651123413147768252411
49309444717696528275628519673751439535754247909
32192066418830117871691225524210700507090646743
82870851449950256586194461543183511379849133691
77992812743384043154923685552678359637410210533
15460313537253257486369091597786903282664591829
83815230286936572873691422648131291743762136325
73032164528297948686257624536221801767322494056
76428193600787207138370723553054463561539464011
85348493792719514594505508232749221605848912910
94518995994868619954314766693801303717616359259
44797461642200508850794698044871332051331607391
34230540198872570038329801246050197013467397175
90902738949392381731578699684589979478106804282
24360937839463352654228157043028324423855150823
16490967285712171708123232790481817268327510112
74678231741098588868370852200071173349225391332
23007561471804290075276777933523062006182860124

Lewis Carroll's Cats and Rats

5525424306100689480544658470482065098266431936 0
9603887362585107470743406362869765767026992586 4
9953557976318173902550891331223294743930343956 1
6132833407283166349825814522686200430779908468 8
1038041873683248009038735962129196336025831207 8
1673673742533322879296907205490595621406888825 9
9124458184237959786347648431567376092362509037 1
5117989414242622702200662864868678687101829808 7
2802560693101949280830825044198424796792058908 8
1711232719230145558291674679519743054802640464 6
8540027339938607985944659615017525869658114475 6
8510041568687730903712482535343839285397598749 4
5849705003822501248928400182659005625128618762 9
9380444073401423470620557853053250349181895897 0
7199305662188512963187501743535960282201038211 6
1604854512103931331225633226076643623668829685 0
2088394961428304847391139916696226499485636852 3
4712873294796680884509405893951104650944137909 5
0227654565313301867063352132302846051943438139 9
8105614006525953007317907727110657834941746426 8
4720956134647327748584238274899668755052504394 2
1823219135722305406671537337424854364566378204 5
7016545932181540535483936142506644985854033074 6
6468541890148134347714650315037954175778622811 7
7658587694168090820312 5

If that was hard for you to grasp, try the following number:

$$3 \uparrow\uparrow\uparrow 4 = 3 \uparrow\uparrow 3 \uparrow\uparrow 3 \uparrow\uparrow 3 = 3 \uparrow\uparrow 3 \uparrow\uparrow 3^{3^3} = 3 \uparrow\uparrow 3 \uparrow\uparrow 3^{27}$$
$$= 3 \uparrow\uparrow 3 \uparrow\uparrow 7,625,597,484,987$$

and forgive me for not writing explicitly the 7,625,597,484,987 exponents of 3 and continuing to evaluate the expression.

Donald Knuth is a truly remarkable person. He is one of the world's greatest computer scientists, a recipient of the prestigious

Turing award for mathematicians and many other prizes. He is the author of the "bible" of computer science, *The Art of Computer Programming*, and is an expert on the complexity analysis of algorithms. Almost every scientist who writes a scientific article uses *Tex*, a word processing language especially designed for writing equations, invented by Knuth. He has also invented many programming languages — indeed he created a new concept of programming called literate programming which in turn contributed to the invention of the popular C programming language. What really stands out is his unusual way of looking at things, his exceptional capability for solving problems in original ways, and his perception of math and computer science as literature. This is probably why he deals a lot with notations of mathematical operations, computer coding and text processing. Knuth, born in 1938, has had an exceptional career, first as an assistant professor at the California Institute of Technology, then at the Institute for Defense Analyses, and from 1969 onwards as Fletcher Jones Professor of Computer Science at Stanford University. To top all that, he is also an accomplished composer and organist, and a recreational mathematician and linguist.

Getting back to operator puzzles, frankly I am not aware of any long history, certainly not of the kind that we posed for you above. I have seen in recent years similar puzzles, though, that go under the generic name "operator search puzzles", or sometimes just "fill in the blanks". Mostly used by teachers to get their students to practice arithmetic, a relatively common challenge is a row of numbers followed by an equal sign and an answer. Between every two numbers you have to place one of the four, elementary operations: $+, -, \times, \div$, so that the equation holds. Here is an example of such a puzzle and its solution:

Puzzle: 2 5 7 10 = 39
Solution: $2 + 5 \times 7 - 10 = 39$

The standard procedure is to insert and apply the operators from left to right as you go, *ignoring* operator precedence rules. In the example, first do 2 + 5 = 7, then multiply by 7 to get 49 and then subtract 10 to get 39. However, thinking about the order of operations gives both puzzle makers and puzzle solvers quite a headache. A good puzzler would make sure operator precedence is stated quite clearly in the instructions, or better still, allow the solver to add parentheses.

Operator precedence rules are very important and taught in school, much to the dislike of young kids who always seem to forget some of them and thus end up with the wrong answers. The rules are straightforward enough: always apply the operators within the brackets first, then calculate exponents, and then perform multiplication and division before addition and subtraction (with no preference between multiplication and division, or addition and subtraction). Operator precedence rules are known by their acronyms as:

- BODMAS (Brackets, Order, Division–Multiplication, Addition–Subtraction) in England and British affiliated territories, and

- PEMDAS (Parentheses, Exponents, Multiplication–Division, Addition–Subtraction) in the United States.

Of course, as more and more operators are discovered, the order of operator precedence has to be updated. We've already seen one example of that above in Knuth's notation. When using exponents of exponents they should be calculated top down, so, 2^{3^2} is equal to $2^9 = 512$, not $(2^3)^2 = 8^2 = 64$.

Polish notation is very interesting. Invented in 1924 by Polish logician Jan Łukasiewicze, it is very useful for computer language interpreters and is employed in the LISP computer language as well as in the image language of PostScript for .ps files. Its advantage is that there is no need to remember the ordering of the operators or the use of parentheses. This is because the operator is written either before (in prefix Polish notation) or after (in postfix or reverse Polish notation) the numbers. As long as all the operators in the expression are of the same kind (called *arity*), i.e. all binary or all unary, then there is no ambiguity when an expression is evaluated. The prefix Polish notation for $4 - 1$ is

$$- \ 4 \ 1$$

and the expression for $(4 - 1) \times 5 = 15$ is

$$\times - 4 \ 1 \ 5 = 15$$

because the operations are evaluated from the innermost out, and from left to right once an operator is followed by two numbers. To complete the example, we note that the Polish prefix expressions for $4 - (1 \times 5) = -1$ and for $(1 \times 5) - 4 = 1$ are

$$- \ 4 \times 1 \ 5 = -1, \quad \text{and} \quad - \times 1 \ 5 \ 4 = 1 \quad \text{respectively.}$$

Another form of operator puzzles (also very recent, though I don't know who was the first to invent these) is the "fill in the blanks" grid, sometimes known as number grid or number square, like the puzzle and its solution shown in Fig. 4.2. These puzzles come in all shapes and sizes, not just grids, and can be automatically generated online on various sites, like Discovery Education's Puzzlemaker.

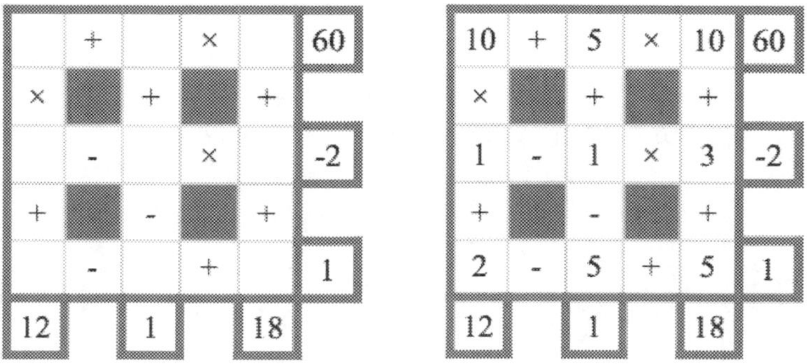

Fig. 4.2 A number grid puzzle (left) and its solution (right)

KenKen™ puzzles, also known as CalcuDoku or MathDoku is a popular puzzle, especially in the West, and was invented in 2004. In its advanced version, the puzzler has to work out not only the correct missing numbers, but also the missing operators. The biography of KenKen™'s inventor, Japanese puzzle master Tetsuya Miyamoto, is inspirational and you can read all about it in Alex Bellos's wonderful book on Japanese puzzles, *Puzzle Ninja*. In a KenKen™ grid, each number appears exactly once in every row and column. The numbers used are the whole numbers between 1 and the number of cells in each row and column. The grid is also divided into "cages". The numbers in each cage combined with one arithmetic operator have to equal the target number written in the top left corner of the first cell in the cage. Only one operator is allowed for all the numbers in each cage. For subtraction and division, any order of operation is allowed. The simplest version of KenKen™ simply uses addition for the whole grid. Most versions give the operator for each cage next to the target number. There are versions that write all the operators that are used outside the grid, so that you know what operators are used, but not where. The advanced puzzles, however, leave the operators out entirely and that's where the fun begins! Figures 4.3, 4.4 and 4.5 show examples of the different versions of KenKen™ puzzles, and their solutions.

There is one operator puzzle that does have a bit of history — the "four fours" puzzle. The aim of the puzzle is to reach a target number using the number 4 four times and any operators that you like. For example, reach the target number 15, using four 4's and any legal arithmetic operators. Here is one possible solution:

$$4 \times 4 - 4 \div 4 = 15.$$

Walter Rouse Ball was one of the first people to recognize recreational math as a genre of its own. He introduced the four fours puzzle in a book, *Mathematical Recreations and Essays*, that he published in 1892. This book could be considered the "bible" of

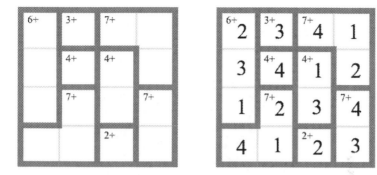

Fig. 4.3   An addition-only KenKen™ puzzle (left) and its solution (right)

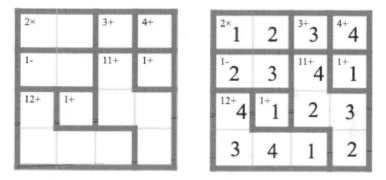

Fig. 4.4   A KenKen™ puzzle with multiple operators (left) and its solution (right)

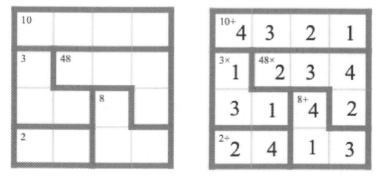

Fig. 4.5   An advanced KenKen™ puzzle with no operators shown (left) and one possible solution (right)

Fig. 4.6   Walter Rouse Ball

recreational math and remains a popular book to this day. Ball wrote the book jointly with British–Canadian mathematician and musician, Harold Scott MacDonald Coxeter, who later became a famous geometer and collaborated with the famous Dutch artist, Maurits Cornelis Escher, on the mathematical aspects of his work. Rouse Ball trained as a mathematician and a lawyer, a practice he gave up when he was appointed lecturer in math at Cambridge University's Trinity College. He was not very active in research, but he was known as a great teacher, chess player, math historian and magician, and most of the works that he published were on these topics. Apart from this, he also took up many administrative positions at Cambridge. Four fours puzzles are really addictive and many generalizations have been suggested. One that is typically useful, because it greatly broadens the scope of the puzzle, is to allow concatenation. Using this, you can get 111 in the following way:

$$444 \div 4 = 111.$$

Of course, there is no reason to limit the puzzle to four 4's. You can make puzzles with five 5's or four 3's or even an assorted mix of numbers.

A similar operator puzzle was formulated by American mathematician, Erich Friedman. The puzzle is to find *Friedman numbers*, numbers that can be calculated using their *own* digits, any regular arithmetic operator and/or concatenation, and discover *how* they are constructed. For example, 126 is a Friedman number because you can calculate it using its own digits, 1, 2 and 6, concatenation, and multiplication:

$$21 \times 6 = 126.$$

Incidentally, a *nice* Friedman number is a Friedman number where the digits in the expression are arranged in the same order as they are in the number itself, for example, the number 343:

$$(3 + 4)^3 = 343.$$

Vampire numbers are a "special case" of Friedman numbers. They are Friedman numbers where the only operations allowed are multiplication and concatenation. The multiplicands are known as *fangs*. 210 and 8700 are the two fangs of the Vampire number 1,827,000:

$$210 \times 8700 = 1827000.$$

If the fangs have the same number of digits, and at least one of them does not end in zero, the number is called a *true* Vampire number, and if the multiplicands are also prime numbers, then the number is a *prime* Vampire number:

$$167 \times 701 = 117067.$$

Clifford Pickover, the inventor of Vampire numbers, first introduced them in an article in *Discover* magazine, "Interview with a number", where he explains the unusual name "vampire" because the numbers seemingly innocently lurk within the whole numbers, and because they can be decomposed into "fangs". Pickover trained as a biochemist and biophysicist, and worked for many years as a computer scientist in IBM's Thomas J. Watson Research Center in New York. He is a prolific recreational math and popular science author who has written over 40 books and over 350 articles, many of them translated into many languages. His writing often celebrates the connections between art, science, math and philosophy.

If you're interested, and by now you should be, in unusual whole numbers, the place to look for many hours of pleasure is *The On-line Encyclopedia of Integer Sequences*, which we briefly mentioned in the previous chapters. The OEIS, as it's commonly known, is a website maintained by Neil Sloane, an American mathematician, who worked at the labs of AT&T. Sloane was born in England, grew up in Australia, and studied math and engineering, first in Australia and later on at Cornell University in the United States where he became an expert in cryptography. In 1965, Sloane, who was then interested in combinatorics, decided to collect and arrange groups of integers — encyclopedia style. So, he started *The Encyclopedia of Integer Sequences* and published two books in 1973 and 1995. When mathematicians started sending him more and more sequences, he realized that a book would not be practical, so he started an online site under the name *The On-line Encyclopedia of Integer Sequences* (OEIS for short). The site became very popular, and as of August 2020, contains over 360,000 sequences. It is perhaps one of the first examples of user-made content, pre-dating Wikipedia by about 5 years. There are many strange and original sequences in the encyclopedia. Amongst the strangest is the sequence 2, 3, 6, 7, 8, 12, 13, 16, 17, 18, 30, 31, ... — the sequence of all the numbers, whose name in Hebrew begins with the letter ש (Shin — phonetically, shin)!

Which brings me back to the operator puzzles I posed for you at the beginning of the chapter. It's fair to say, that these puzzles, although original, build on a foundation of at least one hundred years of similar puzzles and much related history.

Generalizations of the Puzzle

We've already seen one way to generalize operator puzzles — using made-up operators. The last puzzle at the beginning of the chapter is the made-up operator, "concatenate the second number with the first". Making up operators leads to an infinite number of possible operator puzzles. Four fours puzzles can be generalized to other numbers. All of these operator puzzles are part and parcel of what's known in math as number theory, because we are looking for interesting connections between integers.

However, there's much more to operators than that. Indeed, if you asked a mathematician what an operator is, his first choice probably wouldn't be arithmetic operators. Rather, operators have become extremely valuable with respect to mathematical functions. Operating on a function does something to that function, for example, multiply the function by 5, or square the function. These operations on the function x^3, return $5x^3$ and x^6, respectively.

Operations on functions play a key part in quantum mechanics, and one good example is the famous (time-independent) Schrödinger equation:

$$\hat{H}\psi = E\psi.$$

It's amazing how such a simple-looking equation, never mind the unusual symbols, holds the secrets of the Universe. Quantum mechanics is truly a wonderful, mathematical interpretation of the physical world. An astounding human accomplishment. Obviously, we cannot summarize quantum mechanics in one paragraph, and that's not the point anyway — but I would like to show you something I find very neat. The Schrödinger equation is really a puzzle — on unknown functions — and here's why. The (Greek) ψ symbol in the equation symbolizes an (as yet) unknown function, in the same way that an x symbolizes an unknown number in algebra. The E symbolizes a constant — a number, not a function — which is, as yet, also unknown. The H with the funny hat is a well-known operator on functions, called the Hamiltonian (hence the H) — we'll get to its exact definition in a moment, but for now it suffices to say that the Hamiltonian is an operator that, when operated on a function, sums the function's second derivative and its product with another given function. If you're unfamiliar with derivatives, usually learnt in high school, you can either search the internet for a simple explanation, or skip the next paragraph.

Schrödinger's puzzle asks the following: find a function which when you take its second derivative and add that to what you get when you multiply it by another given function, you get back the same function multiplied by a constant. Solve the puzzle, and you will have found one of the secrets of the Universe. You're probably asking how does this math relate to real life? What the operator represents in real life is a measurement. The Hamiltonian operator measures the energy of the system and it has two components: one for kinetic energy and the other for potential energy. For completion, here is one simplified definition of such a Hamiltonian in one dimension:

$$\hat{H} = \frac{d^2}{dx^2} + V(x).$$

The first term in the expression means: take the second derivative and the second term denotes the potential energy. Suppose we are interested in functions that describe the energy of a free particle, a system with no potential energy. We set the second term to zero, and now ask the question: what function ψ, when we take its second derivative will give us back *the same* function multiplied by a constant, and here starts a guessing game — a puzzle! One possible solution is the exponential function e^x, because

$$\frac{d^2}{dx^2}e^x = e^x,$$

meaning that when we apply the Hamiltonian to the exponential function, we get back the function times a constant — which happens to be 1. Two other possible solutions are the trigonometric functions sine and cosine that return the negatives of themselves when we take their second derivatives. In both cases, E, the constant, is equal to -1. Solving the Schrödinger equation for different systems provides us with descriptions of all non-relativistic quantum systems, providing the correct physics for atoms, molecules and many other systems. A generalization of the operator puzzle at the beginning of the chapter to functions can be to find operators that map one function onto another — just like the Schrödinger equation looks for functions that map onto themselves. You can find one such puzzle in the Challenge Yourself! section at the end of this chapter.

Operators are very important mathematical tools partly because they bridge the gap between things you do in the real world and math. Apart from functions, operators can operate on other things as well, like text, or images. The rotation operator, for example, acts on an image, rotating it by a certain angle. There are many ways to mathematically represent images, and there are many ways to represent operators as well. We've already seen that operators can be functions that operate on functions. In the case of images, it turns out that a good representation is by using *matrices*, an arrangement of numbers in rows and columns, since images can be broken down into pixels, each assigned a number that specifies its individual color. Operators can also operate on text, as is used in many text manipulating and spreadsheet computer programs.

Common operators are concatenation, string length calculations and letter substitution.

Yet another type of operators, also found in spreadsheets, are logic operators that operate on statements and specify whether they are "TRUE" or "FALSE". "TRUE" is represented by the number 1 and "FALSE" by a 0. The binary numbers 0 and 1 are also known as bits. Everything in computers is translated into bits, the language that computers understand. This is because these are translated into the physical world of electronics, where 1 signals an electric current in a circuit and 0 its absence.

The "AND" operator is a binary operator that acts on two operands. It returns "TRUE", or "1", only if both statements are true. The "OR" operator is a binary operator that returns "TRUE", or "1", if at least one of the statements is true. The "NOT" operator is a unary logic operator that acts on one operand, returning the opposite of its logical statement. Here are some examples of these operators at work:

$$12 > 3 \text{ AND } 4 < 12 = \text{"TRUE"} = 1$$
$$12 > 3 \text{ AND } 4 < 0 = \text{"FALSE"} = 0$$
$$12 > 3 \text{ OR } 4 < 0 = \text{"TRUE"} = 1$$
$$12 > 47 \text{ OR } 4 < 0 = \text{"FALSE"} = 0$$
$$\text{NOT } 12 > 3 = \text{"FALSE"} = 0$$
$$\text{NOT } 12 > 43 = \text{"TRUE"} = 1$$

One of the questions that I often ask myself when I'm trying to solve a puzzle is what are the mental processes that are at work? With operator puzzles it seems at first that it's just a guessing game, but a closer analysis suggests two methodologies that are unconsciously at work. One is trial and error. Guessing is random. Trial and error on the other hand is a learning process. Although your initial choice, like "let's try addition" can be random, if this is a wrong choice, you've still gained insight to the problem and your following choices will build on that. Another process at work is logic.

Logical reasoning restricts your choices and, in many cases, dictate a unique solution. At the beginning of the chapter we mentioned cryptarithms, arithmetic equations where the operators are given, but the numbers are encrypted. Logical reasoning is particularly useful for solving cryptarithms. For example, suppose you are trying to solve the following cryptarithm: IN + AN = INN. Notice the units column: $N + N = N \mod (10)$. Logical reasoning dictates $N = 0$ because no other digit equals twice itself. It also dictates that $I = 1$, because the sum of two 2-digit numbers can never exceed 198. The solution, based on logic alone, is unique and immediate: $N = 0$, $I = 1$, $A = 9$ and $10 + 90 = 100$. Similar reasoning can be used for more advanced problems.

Recap

In this chapter we tackled the concept of operators and operator puzzles. We saw a few different types of operators: arithmetic operators and operators that act on functions, texts, images and logical statements in the form of bit operators. Operator puzzles are mathematical expressions where the operator has been omitted or encrypted, and the puzzle solver has to try and discover the missing operator or operators. We saw a few varieties of operator puzzles, from Rouse Ball's four fours puzzles, Erich Friedman's Friedman numbers and Clifford Pickover's Vampire numbers, through operator grid puzzles and KenKen™ puzzles, to the original operator puzzles we posed at the beginning of the chapter. Along the way we learnt about the history of mathematical symbols, discussed BODMAS and PEMDAS preference rules for operators and learnt about Polish notation. Donald Knuth's up-arrow notation took us into the world of contemporary notation — operator design for modern math. We also saw how quantum mechanics uses operators and how operators connect math to real-life phenomena.

Here are some short, informal definitions of some of the terms in this chapter:

- *Binary operators* — operators that act upon two operands.

- *BODMAS/PEMDAS* — acronyms for the order of arithmetic operations in an expression:
  — British: Brackets, Order, Division–Multiplication, Addition–Subtraction.
  — US: Parentheses, Exponents, Multiplication–Division, Addition–Subtraction

- *Cryptarithms* — puzzles where digits in an arithmetic expression are encrypted and the aim is to decipher them.

- *Four fours* — Walter Rouse Ball's puzzle, whose aim is to reach a target number using the number 4 exactly four times and any arithmetic operators of your choice.

- *Friedman number* — A number that can be calculated using its own digits, any regular arithmetic operator and/or concatenation.

- *Grid operator puzzles, KenKen™ puzzles, CalcuDoku, MathDoku* — various kinds of grid puzzles.

- *Nice Friedman number* — A Friedman number where the digits used in the construction appear in the same order as in the number itself.

- *OEIS* — *The On-line Encyclopedia of Integer Sequences*, created and maintained by Neil Sloane.

- *Operator* — a mathematical symbol that typically represents an *operation,* an action performed on one, two or more *operands.* The definition might be easier to understand with a few examples of operators and some of their operands:
  — the set of arithmetic operators: $+$, $-$, $\times$, $\div$, etc. that operate on numbers.

— the set of textual operators: *length of text, concatenation, locations of certain letters in text,* etc. that operate on strings of text.

— the set of logical operators: AND, OR, NOT, XOR, etc. that operate on logical statements.

— the set of mapping operators: *taking the derivative, integration,* etc. that operate on functions.

- *Operator puzzles* — puzzles where the aim is to find missing operators.

- *Polish notation* — A notation where the operators are written either before (prefix) or after (postfix) the numbers upon which they act.

- *Prime Vampire number* — A Vampire number where the fangs are prime numbers.

- *Schrödinger equations* — Two main equations, one known as the time-dependent equation and the other as the time-independent equation, that govern the structure and dynamics of isolated quantum systems.

- *Ternary operators* — operators that act upon three operands.

- *True Vampire number* — A Vampire number where the fangs have the same number of digits, and at least one of them does not end in zero.

- *Unary operators* — operators that act upon one operand.

- *Up-arrow notation* — "laddered" exponentiation, where the number of arrows denote an increasing degree of exponentiation. So if the number preceding the arrows is the base a and the number after the arrows is the power b, then $a \uparrow b$ is equivalent to performing the operation 3 <u>multiplied</u> by itself b times, and $a \uparrow\uparrow b$ is equivalent to performing the operation a <u>exponentiated</u> by itself b times, etc.

- *Vampire number* — A number that can be reached by multiplying numbers, known as *fangs* that are constructed using only the number's own digits.

Challenge Yourself!

1. Find the missing operators in the following puzzles. The missing operators are always denoted by a question mark (?), and are all well-known arithmetic operators.

(a) $12 ? 4 = 3$
$$1 ? 1 = 1$$
$$0 ? 4 = 0$$
$$140 ? 14 = 10$$

(b) $? 0 = 0$
$$? 1 = 1$$
$$? 12 = 144$$
$$? 5 = 25$$

(c) $? 5 = 120$
$$? 1 = 1$$
$$3 = 6$$
$$? 8 = 40320$$

(d) $? 2 = 0.5$
$$? 1 = 1$$
$$? 0.5 = 2$$
$$? 10 = 0.1$$

2. Find the missing *made-up* operators in the following puzzles. The missing operators are always denoted by a question mark (?).

   (a) $? \, 0 = 3$
         $? \, 7 = 10$
         $? \, 112 = 115$
         $? \, (-2) = 1$
         $? \, 100 = 103$

   (b) $? \, 0 = 0$
         $? \, 8 = 8$
         $? \, 10 = 1$
         $? \, 132 = 6$
         $? \, 55 = 10$
         $? \, 641 = 11$

   (c) $12 \, ? \, 2 = 140$
         $1 \, ? \, 1 = 0$
         $0 \, ? \, 4 = -16$
         $13 \, ? \, 12 = 25$
         $-2 \, ? \, 0 = 4$

   (d) $9 \, ? \, 4 = 5$
         $1 \, ? \, 1 = 2$
         $100 \, ? \, 64 = 18$
         $0 \, ? \, 0.25 = 0.5$
         $4 \, ? \, 4 = 4$

3. Find the missing *text* operators in the following puzzles. The missing operators are always denoted by a question mark (?).

   (a) ? "This is a random sentence" = 25

   ? "Hello!" = 6

   ? "My name is David" = 16

   (b)   "Hello" ? " World!" = "Hello World!"

   "Yes" ? "No" = "YesNo"

   "1" ? "2" = "12"

   (c)   ? "I love playing cards!" = "p"

   ? "Charlotte" = "t"

   ? "Great Britain" = "r"

4. Find the missing *logic* operators in the following puzzles. The missing operators are always denoted by a question mark (?). Reminder: The truth value of "TRUE" is 1 and the truth value of "FALSE" is 0.

   (a)   ? 1 = 0

   ? "All cats are black" = 1

   ? (17 = 16) = 1

   ? 10 = 01

   (b)   1 ? 1 = 1

   0 ? 1 = 0

   1 ? 0 = 0

   0 ? 0 = 0

   (c)   1 ? 1 = 1

   0 ? 1 = 1

   1 ? 0 = 1

   0 ? 0 = 0

5. Evaluate the following expressions:

   (a) $+ \times 7\ 3\ 12 =$

   (b) $\times - + \div 12\ 2\ 65\ 18\ 23 =$

   (c) $7 \uparrow 8 =$

   (d) $2 \uparrow \uparrow 4 =$

   (e) $1 \uparrow \uparrow \uparrow 3 =$

   (f) $2 \uparrow \uparrow \uparrow \uparrow 2 =$

6. Calculate all the target whole numbers from 0 to 20, using four 4's and any known arithmetic operation, including squaring the number, taking the square root and factorial. Concatenation is allowed. Show how the calculation is done!

7. Solve the following number grid and CalcuDoku puzzles using positive whole numbers less than 10. For the number grid, the operator ordering should be according to BODMAS/PEMDAS preference rules for operators, unless the operator is the same throughout the row/column, in which case the ordering is from left to right and top to bottom. For the CalcuDoku puzzles, subtraction and division within a cage can be performed in any order of operation.

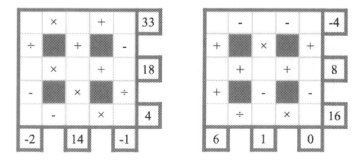

Fig. 4.7 Two number grid puzzles

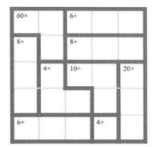 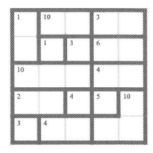

Fig. 4.8   Two CalcuDoku puzzles. The operators that are omitted in the puzzle on the right, can be addition, subtraction, multiplication or division.

8. Solve the following cryptarithms:

$$MA + PA = ARM$$
$$HAND + BAND = HEADS$$
$$CALL + THE = INLET$$

9. The question mark in the following problem represents a missing operator that is a function that maps the first function in the following expressions onto the second function. The missing operator (function) is the same for all the expressions. What is it?

$$? \, 1 = x$$
$$? \, x = \tfrac{1}{2} \, x^2$$
$$? \, \cos(x) = \sin(x)$$
$$? \, e^x = e^x$$

Solutions

1. (a) The missing operator is $\div$:

$$12 \div 4 = 3$$
$$1 \div 1 = 1$$
$$0 \div 4 = 0$$
$$140 \div 14 = 40320$$

(b) The missing operator is ^{2} (squaring):

$$0^2 = 0$$
$$1^2 = 1$$
$$12^2 = 144$$
$$5^2 = 25$$

(c) The missing operator is ! (factorial):

$$5! = 120$$
$$1! = 1$$
$$3! = 6$$
$$8! = 40320$$

(d) The missing operator is $\frac{1}{x}$ (the reciprocal):

$$\frac{1}{2} = 0.5$$
$$\frac{1}{1} = 1$$
$$\frac{1}{0.5} = \frac{1}{\frac{1}{2}} = 2$$
$$\frac{1}{10} = 0.1$$

2. (a) The missing operator is "add 3 to the number":

$$3 + 0 = 3$$
$$3 + 7 = 10$$
$$3 + 112 = 115$$
$$3 + (-2) = 1$$
$$3 + 100 = 103$$

(b) The missing operator is "the sum of the digits of the number":

the sum of the digits of $0 = 0$

the sum of the digits of $8 = 8$

the sum of the digits of $10 = 1$

the sum of the digits of $132 = 6$

the sum of the digits of $55 = 10$

the sum of the digits of $641 = 11$

(c) The missing operator is "the square of the first number minus the square of the second number":

$$12^2 - 2^2 = 140$$
$$1^2 - 1^2 = 0$$
$$0^2 - 4^2 = -16$$
$$13^2 - 12^2 = 25$$
$$(-2)^2 - 0^2 = 4$$

(d) The missing operator is "the square root of the first number plus the square root of the second number":

$$\sqrt{9} + \sqrt{4} = 5$$
$$\sqrt{1} + \sqrt{1} = 2$$
$$\sqrt{100} + \sqrt{64} = 18$$
$$\sqrt{0} + \sqrt{0.25} = 0.5$$
$$\sqrt{4} + \sqrt{4} = 4$$

3. (a) The missing *text* operator is "the number of characters":

The number of characters in "This is a random sentence" $= 25$

The number of characters in "Hello!" $= 6$

The number of characters in "My name is David" $= 16$

(b) The missing *text* operator is "concatenation" (denoted by +):

"Hello" + " World!" = "Hello World!"

"Yes" + "No" = "YesNo"

"1" + "2" = "12"

(c) The missing *text* operator is "the eighth character":

The eighth character in "I love playing cards!" = "p"

The eighth character in "Charlotte" = "t"

The eighth character in "Great Britain" = "r"

4. (a) The missing *logic* operator is "NOT":

$$\text{NOT } (1) = 0$$
$$\text{NOT (All cats are black)} = 1$$
$$\text{NOT } (17 = 16) = 1$$
$$\text{NOT } (10) = 01$$

(b) The missing *logic* operator is "AND":

$$1 \text{ AND } 1 = 1$$
$$0 \text{ AND } 1 = 0$$
$$1 \text{ AND } 0 = 0$$
$$0 \text{ AND } 0 = 0$$

(c) The missing *logic* operator is "OR":

$$1 \text{ OR } 1 = 1$$
$$0 \text{ OR } 1 = 1$$
$$1 \text{ OR } 0 = 1$$
$$0 \text{ OR } 0 = 0$$

5. (a) $+ \times 7\ 3\ 12 = (7 \times 3) + 12 = 33$

(b) $\times - + \div 12\ 2\ 65\ 18\ 23 = 23 \times (18 - (65 + (12 \div 2))) = -1219$

(c) $7 \uparrow 8 = 7^8 = 5764801$

(d) $2 \uparrow \uparrow 4 = 2^{2^{2^2}} = 2^{2^4} = 2^{16} = 65536$

(e) $1 \uparrow \uparrow \uparrow 3 = 1 \uparrow \uparrow 1 \uparrow \uparrow 1 = 1 \uparrow \uparrow 1 = 1 \uparrow 1 = 1$

(f) $2 \uparrow \uparrow \uparrow \uparrow 2 = 2 \uparrow \uparrow \uparrow 2 = 2 \uparrow \uparrow 2 = 2^2 = 4$

6. $4+4-4-4=0$

$4 \div 4 + 4 - 4 = 1$

$4 \div 4 + 4 \div 4 = 2$

$(4+4+4) \div 4 = 3$

$\sqrt{4} \times \sqrt{4} + 4 - 4 = 4$

$\sqrt{4} \times \sqrt{4} + 4 \div 4 = 5$

$\sqrt{4} + 4 + 4 - 4 = 6$

$4 + 4 - 4 \div 4 = 7$

$4 \times 4 - 4 - 4 = 8$

$4 + 4 + 4 \div 4 = 9$

$(44 - 4) \div 4 = 10$

$44 \div (\sqrt{4} \times \sqrt{4}) = 11$

$4 \times 4 - \sqrt{4} - \sqrt{4} = 12$

$44 \div 4 + \sqrt{4} = 13$

$4 + 4 + 4 + \sqrt{4} = 14$

$4 \times 4 - 4 \div 4 = 15$

$4 \times 4 - 4 + 4 = 16$

$4 \times 4 + 4 \div 4 = 17$

$4! - (4 + 4 - \sqrt{4}) = 18$

$4! - (4 + 4 \div 4) = 19$

$4! - 4 + 4 - 4 = 20$

7.

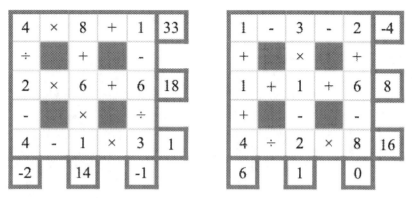

Fig. 4.9 Solutions to two number grid puzzles

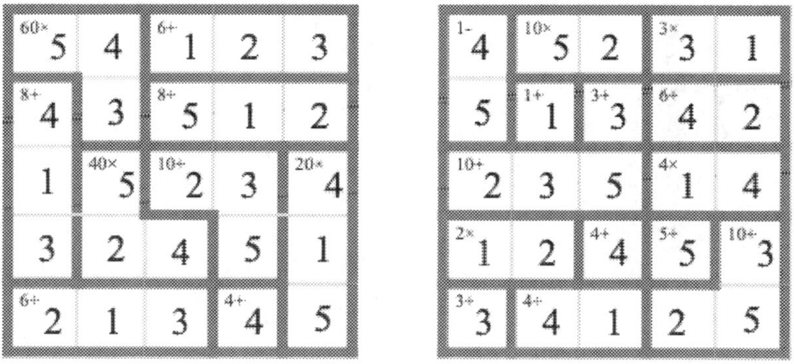

Fig. 4.10 Solutions to two CalcuDoku puzzles

8. MA + PA = ARM $\qquad$ 21 + 81 = 102

  HAND + BAND = HEADS $\qquad$ 1962 + 8962 = 10924

  CALL + THE = INLET $\qquad$ 9744 + 682 = 10426

9. The missing operator is indefinite integration ($\int dx$):

$$\int 1 \, dx = x$$
$$\int x \, dx = \tfrac{1}{2} x^2$$
$$\int \cos(x) \, dx = \sin(x)$$
$$\int e^x \, dx = e^x$$

Bibliography and Further Reading

Friedman, E., (2015). Friedman numbers: can be written in a nontrivial way using their digits and the operations + − * / ^ and concatenation of digits (but not of results), *The On-line Encyclopedia of Integer Sequences.* Retrieved December 26, 2020, from https://oeis.org/A036057

Fuhrer, R. (Ed.) *Book Series: KenKen: Math & Logic Puzzles That Will Make You Smarter!* World Scientific, NJ.

Griffiths, D. J. (2005). *Introduction to Quantum Mechanics.* Pearson Prentice Hall, Upper Saddle River, NJ.

Hadar, E., (2002). Numbers whose Hebrew name begin with the letter "Shin", *The On-line Encyclopedia of Integer Sequences.* Retrieved December 26, 2020, from https://oeis.org/A072426

Miller, J. (2017). Earliest Uses of Symbols of Operation. Retrieved December 26, 2020, from http://jeff560.tripod.com/operation.html

Miyamoto, T. (2020). KenKen Puzzles That Make You Smarter. http://www.kenkenpuzzle.com/

Pickover, C. (1995). Interview with a number, *Discover.* June 16(6).

Strachan, L. (2014). The Friedman numbers, *Numbers Are Forever.* Hachette UK, p. 70.

Wilson, D. W., (2009). Vampire numbers: (definition 1): *n* has a nontrivial factorization using *n*'s digits, *The On-line Encyclopedia of Integer Sequences*. Retrieved December 26, 2020, from https://oeis.org/A020342

From Number Pyramids to Fractals

Introduction

The following puzzle is usually performed as a very entertaining magic trick. However, its mathematical background is truly overwhelming — a great excuse for us to venture into Pascal's triangle and the world of fractals. There are a few versions to this puzzle or trick, and numbers or playing cards can be used to create a nice effect. Here is my own version.

The Puzzle

Eleven random, whole numbers are written in a line. Every two adjacent numbers are summed, and the units digit of the sum is retained and written in a line above and in between the two addends. There are now ten whole numbers in the new line. This process is repeated with the new line and continued until there is a single number, which will be at the apex of a "number pyramid".

Predict (mentally) what this apex number will be <u>without</u> actually building the pyramid if the initial number line is

12 3 55 9 0 21 87 4 1 12 8.

Where to Start?

This puzzle is a hard one, if you strictly obey the rules. However, things get simpler since you *can* actually calculate the correct answer by building the pyramid, and then you'll know the target number you have to aim for. Once you've done that, you can try to spot a connection between that number and the numbers in the initial line. A more methodological method, indeed the one I used on another version of this puzzle, is to reduce the puzzle to a much simpler one (perhaps just an initial number line of four numbers), replace the numbers with the letters *a*, *b*, *c* and *d*, and use algebra to work out a formula for the apex number. Once you've done that you can generalize for eleven numbers.

Solving the Puzzle

Following the hints in the previous section, I'll use both methods. First let's build the pyramid and see that the apex number is 2:

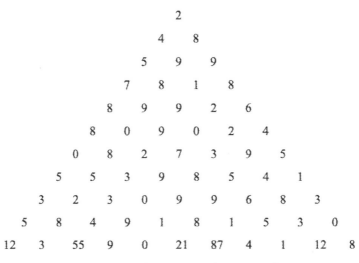

Fig. 5.1   Solution to number pyramid

Trying to find any obvious connection between the numbers in the bottom row and the number 2 is fruitless. There just seems to be too much information. Therefore, we'll try the other suggestion, using algebra, to see if we can get something. First, we'll use a smaller pyramid — just four levels high, substituting the numbers with the letters a, b, c and d and using the appropriate algebra. For the moment, let's also forget about keeping the units digit and just sum regularly. Figure 5.2 shows this algebraic number pyramid.

$$a+3b+3c+d$$

$$a+2b+c \qquad b+2c+d$$

$$a+b \qquad b+c \qquad c+d$$

$$a \qquad b \qquad c \qquad d$$

Fig. 5.2   An algebraic number pyramid

This is a good start. We can see that we can calculate the apex number, using the expression:

$$a + 3b + 3c + d \qquad (5.1)$$

i.e. taking the first number in the line, adding to it three times the second and third numbers, and finally adding the last number, as verified by the example in Fig. 5.3 where the initial line is a random line of numbers. If you don't trust me (or algebra ...) you can build your own random line of numbers and check it out!

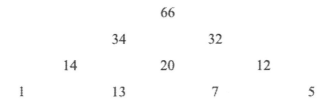

Fig. 5.3   Example of a four-level number pyramid with a random initial line

Examining the pyramid in Fig. 5.3, you can see that, indeed,

$$1 + 3 \times \mathbf{13} + 3 \times \mathbf{7} + \mathbf{5} = \mathbf{66}.$$

Admittedly, it's a bit tedious, but doing the algebra for eleven lines gives you the following expression for the apex number:

$$a + 10b + 45c + 120d + 210e + 252f + 210g + 120h + 45i + 10j + k$$

$$(5.2)$$

You are probably thinking that what I'm suggesting is absurd. It seems just as hard to mentally calculate expression (5.2) as it is to mentally reconstruct the number pyramid! Here's where the special addition rule comes in. Remember that we're only keeping the units digit when we sum, so if that digit happens to be 0, it doesn't contribute anything to the sum. Now take another look at (5.2). Immediately we can see that the units digit of the second, fourth, fifth, seventh, eighth and tenth in the expression are all zero — because the numbers are all multiplied by a multiple of ten! Using similar logic for the numbers that are multiplied by 45, the units digits of the third and ninth numbers in the expression will be either 0 if the number is even, or 5 if the number is odd. Finally, the units digit of the number bang in the middle of the expression will be twice its own unit digit, because this is what you get when you multiply the number by 252 — the first (hundreds) 2 and the middle 5 of 252 do not affect the units digit of the actual product $252 \times number$. If you don't believe me (or math), just try out an example. If the initial line is

$$1\ 3\ 8\ 13\ 5\ 24\ 43\ 5\ 11\ 99\ 7,$$

plugging these numbers into (5.2) gives

$$1 + 30 + 360 + 1560 + 1050 + 6048 + 9030 + 600 + 495 + 990 + 7$$
$$= 20171$$

and taking just the units digit of this gives

$$1 + 0 + 0 + 0 + 0 + 8 + 0 + 0 + 5 + 0 + 7 = 21,$$

whose units digit is indeed 1, the units digit of 20171.

So to calculate the apex number from the original line, here's what you do:

1. Sum the units digit of the first and last numbers in the line.
2. Add twice the units digit of the number in the middle of the line.
3. If the third number is odd, add 5.
4. If the ninth number is odd, add 5.
5. The units digit of the sum is the apex number!

Using this for the line in the puzzle,

$$12 \; 3 \; 55 \; 9 \; 0 \; 21 \; 87 \; 4 \; 1 \; 12 \; 8$$

we get

$$2 + 8 + 2 + 5 + 5 = 22$$

and the units digit of that is 2, which is what we found to be the apex number of the pyramid.

The History of the Puzzle and Related Topics

This version of the number pyramid is one that I presented at a conference, but it has its roots in two very similar number pyramids, that first appeared in recreational math and popular science writer Martin Gardner's book, *Mathematical Carnival*. However, I would like to take you on a journey much further back in time, to discuss two concepts that are very much connected to this puzzle. One of them is the concept of *modular arithmetic* and the other, *Pascal's triangle*.

Fig. 5.4   Portrait of Blaise Pascal, Palace of Versailles

Blaise Pascal was an extraordinary 17th century French mathe-matician, philosopher, physicist and theologian. Home-educated by his father, Etienne, himself an amateur mathematician and professional lawyer, he made some of the most important math and science discoveries that are still very much in use today. Etienne thought it inappropriate for young Pascal to learn math at a young age, so he threw out all the mathematical texts in the house and banned the young boy from learning math. This proved counterproductive as it raised the kid's curiosity and by the age of 12 he was already producing his own mathematical proofs. His father relented, and the two regularly attended and presented at the main French math meetings hosted by a famous monk, philosopher and scientist, Marin Mersenne.

Pascal suffered quite a lot during his lifetime. His mother died when he was only three, and his father's death when he was 28, greatly affected his own life, and brought about a rift between himself and his sister, Jacqueline, whom he reconciled with only later on in his life. Pascal himself also suffered from poor health for nearly all of his life.

Throughout his life, Pascal had a leaning toward Christianity, and he went through a few religious experiences. In 1646, influenced by his father's doctors, he joined a Catholic splinter group that practiced Jansenism. His affiliation with the group lasted only a couple of years but his religious experiences intensified again after his father's death in 1651. When he was just 30, Pascal had a near-death accident when his carriage very nearly fell into the Seine river. Pascal was rescued, but from that moment on, he devoted the rest of his short life to Jansenist Christianity. One of his famous theological ideas, known as *Pascal's wager*, states that, "If God does not exist, one will lose nothing by believing in him, while if he does exist, one will lose everything by not believing." Pascal died from malignant stomach and brain tumors when he was only 39.

Fig. 5.5   The Pascaline

Pascal's contributions to science and math are truly amazing. He worked on conic sections and projective geometry, made major contributions to fluid mechanics and the study of pressure, and laid the foundations for probability theory. The SI (international standard measurement system) unit for pressure, the Pascal, is named after him. The world's second calculator, the Pascaline, was also named after him (the first was invented by Wilhelm Schickard just a few

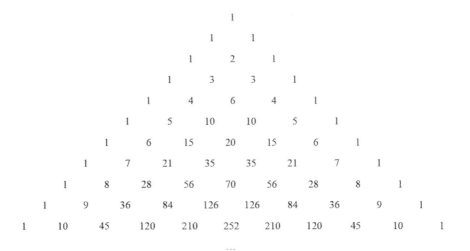

Fig. 5.6   Pascal's triangle

years before). The story goes that Pascal invented the Pascaline, a digital, mechanical, arithmetic calculating machine to help his father in his new job as tax collector. Although fifty prototype machines were manufactured, few were sold.

Which brings us to Pascal's triangle. Pascal's triangle is a number triangle, not unlike the triangle in the number pyramid puzzle. It is usually generated top down, beginning with the number 1 at the apex. Entries in the next row are the sums of the two adjacent numbers directly above it, as shown in Fig. 5.6. If the entries are blank, assume hidden zeros; this explains the 1's at the borders.

It is often thought that Pascal himself invented the triangle, however, the triangle appeared in many places long before Pascal's time. The first figurative description of the triangle is in Indian works from the 10th century CE, and subsequently in Chinese, Persian and Arabic manuscripts. Gersonides, the French–Jewish rabbi, astronomer, physicist, mathematician and philosopher, was the first European to publish a multiplicative formula to calculate what's known as *binomial coefficients* that make up the numbers in the triangle — which we'll get to in just a minute. The first European publication of the triangle in its full figurative manner was by the German mathematician, philosopher and humanist, Peter Bienewitz (or Bennewitz), as the frontispiece of a book that he wrote about business calculations in 1527. So why did the 18th century French mathematician, Pierre Raymond de Montmort, call it Pascal's triangle? Probably because of Pascal's ingenious use of it in combinatorics and probability theory, fields of math where Pascal was one of the first to lay down a modern theory that is still used today.

What makes Pascal's triangle so popular? First, it's a great example of a dynamic number system. Starting with a single number (actually three, including two "hidden" zeros) and a generating function, you create an infinite number triangle. Second, it holds many surprises. Take the diagonals for example. The first diagonal you encounter, either from the left or the right, is a series of ones. Believe it or not, even this series is an interesting mathematical object, called a *repunit*. A repunit is a series of a repetition of ones. 1111 is a repunit of length 4. Which repunits are prime numbers? To date, only five have been proven to be prime. Those of lengths 2 (the number 11), 19 (the number 1,111,111,111,111,111,111) and those of lengths 23, 317 and 1031.

The next diagonal is of course, the *natural numbers*, 1, 2, 3, etc. I always get confused between these and the *whole numbers* which include zero — so I try to remind myself that natural numbers are those used for counting and ordering. You don't count nothing!

The next diagonal, 1, 3, 6, 10, etc. is what's known as the *triangular numbers*. These are *figurate numbers*, numbers that can be arranged in a geometric arrangement. Admittedly, this is a bit of a loose definition, and there is no consensus on what exactly a figurate number is, but at least for *square numbers* like 1, 4, 9, 16, etc. and triangular numbers, the meaning is straightforward

Fig. 5.7   The first four triangular numbers

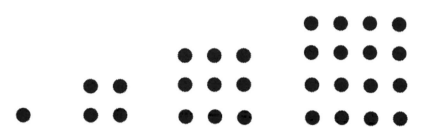

Fig. 5.8   The first four square numbers

and understood once you see the arrangement, as you can see in Figs. 5.7 and 5.8.

It always comes as a surprise to people to find out that there is a literal, visual meaning to square numbers but, historically, square numbers came from the pictorial definition, rather than the abstract n^2. The same goes for triangular numbers. What's more, Fig. 5.9 is a pictorial proof of this cool theorem: "The sum of two consecutive triangular numbers is a square number". This also means that the square numbers also appear in Pascal's triangle as the sum of two consecutive numbers along the triangular number diagonals.

Triangular and square numbers have so many beautiful properties, I could have written a whole chapter on them. One useful property of triangular numbers is that they represent the sum of consecutive whole numbers:

$$1, 1 + 2 = 3, 1 + 2 + 3 = 6, 1 + 2 + 3 + 4 = 10, 1 + 2 + 3 + 4 + 5 = 15,$$

and so on. This is readily understood since a triangular number can be generated from the one preceding it by adding one more line

Fig. 5.9   The sum of two consecutive triangular numbers is a square number

along the long base of the triangle. This line will have exactly one more point than the base of the preceding triangle.

There is a 3-dimensional analog to triangular numbers, known as *tetrahedral numbers*. Imagine that, instead of dots or points, you have small spheres, perhaps oranges. You can arrange them into a triangular pyramid any height you like. This pyramid, that has a triangular base, is also known as a *tetrahedron*. The smallest tetrahedron, after the trivial one orange (a one-layer pyramid) is the two-layer, four-orange tetrahedron. The base will have three oranges and the apex just one. The next pyramid will have ten oranges: six for the base, three in the middle and one on top. The four-layer tetrahedron will add a ten-orange base to the previous one, giving a twenty-orange tetrahedron. The numbers 1, 4, 10, 20 are the *tetrahedral numbers* and, surprisingly, they occupy the

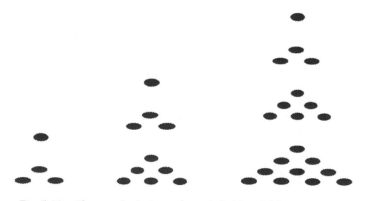

Fig. 5.10   The tetrahedral numbers 1, 4, 10 and 20

diagonal right next to the triangular numbers in Pascal's triangle. Note that each consecutive layer of the triangular pyramid holds the corresponding, consecutive triangular number of oranges, i.e. in the twenty-orange pyramid the layers, from the top down, are 1, 3, 6, 10. By now, you may have guessed that the next diagonal along holds the 4th dimensional equivalent of the triangular/tetrahedral numbers, and indeed it does! The numbers 1, 5, 15, 35, 70, etc. are known as the 4-tetrahedral numbers, where the prefix 4 denotes the dimension. The next diagonals are all triangular/tetrahedral numbers in higher dimensions. Another name for these figurative numbers in different dimensions — and why not call them another name just to make things more complicated — is *simplex numbers*. The 0-simplex is a point, the 1-simplex is a line, the 2-simplex a triangle, the 3-simplex a tetrahedron, and so on.

Pascal's triangle holds many secrets. The *"hockey stick theorem"*, for example, states that if you sum the numbers along any diagonal starting from the number 1, wherever you stop, the sum will always add up to the number in the cell directly below the number you stopped on the opposite diagonal. An example is presented in Fig. 5.11. The numbers along the shaded diagonal, 1 through 70, add up to 126, the number directly below 70 on the opposite diagonal (shaded lighter). Figure 5.11 also shows where the Fibonacci numbers 1, 1, 2, 3, 5, 8, etc. are hidden in Pascal's triangle. The Fibonacci numbers is the series of numbers, where each number in the sequence is the sum of its two preceding numbers. They can be found in Pascal's triangle as the sum of the numbers along the offset diagonals, depicted by the dashed arrows (we've only shown the first six numbers).

What about the rows? Sum the numbers in each row. What do you get? 1, 2, 4, 8, 16, 32, ... the powers of 2! Concatenate the numbers in each row. What do you get? 1, 11, 121, 1331, 14641, ... the powers of 11! Indeed, all rows are powers of 11. The first five rows are evident. From the sixth row onwards, we begin having

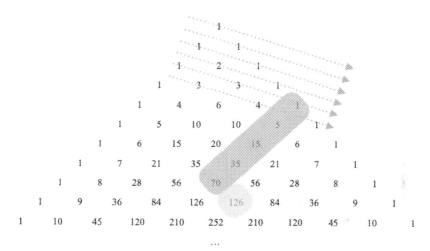

Fig. 5.11   The hockey stick theorem and some of the Fibonacci numbers in Pascal's triangle

two-digit numbers, and in such instances, you have to "carry" over as you would in addition.

Look at the sixth row, 1 5 10 10 5 1. Moving from right to left:

at the first 10, write 0 and "carry" the 1 to the next 10 to get 11
$\rightarrow$ 1 5 ^{1}10 0 5 1
then write 1 and "carry" the 1 to the next digit, 5, to get 6
$\rightarrow$ 1 ^{1}5 1 0 5 1 $\rightarrow$ 1 6 1 0 5 1.
Hence $11^5 = 161051$.

One of the most important properties of the numbers in each row is that they are the *coefficients* in the *binomial expansion*. What does this mean? You have probably learned the formula for squaring the sum of two numbers, a and b:

$$(a + b)^2 = a^2 + 2ab + b^2. \tag{5.3}$$

For example, if we use the numbers 3 and 5 for a and b respectively, we get:

$$(3 + 5)^2 = 3^2 + 2 \times 3 \times 5 + 5^2 = 9 + 30 + 25 = 64,$$

which is indeed, 8 squared. Equation (5.3) is called a binomial expansion. A *coefficient* is a constant number that multiplies an algebraic expression, like the 2 in the expression $2ab$, so the coefficients in Eq. (5.3) are 1, 2 and 1, respectively. Similar formulas were derived for all possible powers. Here are the first seven:

$$(a + b)^0 = 1$$
$$(a + b)^1 = a + b$$
$$(a + b)^2 = a^2 + 2ab + b^2$$
$$(a + b)^3 = a^3 + 3a^2b + 3b^2a + b^3$$
$$(a + b)^4 = a^4 + 4a^3b + 6a^2b^2 + 4b^3a + b^4$$
$$(a + b)^5 = a^5 + 5a^4b + 10a^3b^2 + 10b^3a^2 + 5b^4a + b^5.$$

Look at the coefficients of these expansions, the numbers in bold:

$$\mathbf{1}$$
$$\mathbf{1}a + \mathbf{1}b$$
$$\mathbf{1}a^2 + \mathbf{2}ab + \mathbf{1}b^2$$
$$\mathbf{1}a^3 + \mathbf{3}a^2b + \mathbf{3}b^2a + \mathbf{1}b^3$$
$$\mathbf{1}a^4 + \mathbf{4}a^3b + \mathbf{6}a^2b^2 + \mathbf{4}b^3a + \mathbf{1}b^4$$
$$\mathbf{1}a^5 + \mathbf{5}a^4b + \mathbf{10}a^3b^2 + \mathbf{10}b^3a^2 + \mathbf{5}b^4a + \mathbf{1}b^5.$$

Bingo! We get Pascal's triangle! The *binomial theorem* (another name for the expansion) was known, at least for the first few expansions, in 7th century Indian math, and was tabulated and formulated by Arab, and later on European, mathematicians. Pascal studied these numbers extensively, especially with respect to their use in *combinatorics,* the math of counting things, and probability theory.

A typical combinatorial problem goes like this: suppose there are three kids, John, Matthew and Sarah. How many different ways are there to choose one kid out of the three? Obviously, there are three choices: we can pick John, or Matthew, or Sarah. Now, suppose we want to choose a pair of kids. In how many different ways can we do this? The answer is again, three. We can choose the pair John–Matthew, or Sarah–Matthew, or John–Sarah. Along comes Rachel. Now, there are four different ways to choose one of them, six different ways to choose pairs and four different ways to choose three kids out of the four: John–Matthew–Rachel, or Sarah–Matthew–Rachel, or John–Sarah–Rachel or John–Matthew–Sarah. Now let's add the obvious statements: there is only one way to choose no kid from any group, and there is only one way to choose all kids from a group. Take a look at the numbers again. I'll also add the numbers for a group of five kids.

- For a group of three kids, there is **1** way to choose 0 kids, **3** ways to choose 1 kid, **3** ways to choose 2 kids and **1** way to choose 3 kids.

- For a group of four kids, there is **1** way to choose 0 kids, **4** ways to choose 1 kid, **6** ways to choose 2 kids, **4** ways to choose 3 kids and **1** way to choose 4 kids.

- For a group of five kids, there is **1** way to choose 0 kids, **5** ways to choose 1 kid, **10** ways to choose 2 kids, **10** ways to choose 3 kids, **5** ways to choose 4 kids and **1** way to choose 5 kids.

Take a look at the bold numbers — they are the binomial coefficients — and the third to fifth rows in Pascal's triangle:

$$1\ 3\ 3\ 1, 1\ 4\ 6\ 4\ 1, 1\ 5\ 10\ 10\ 5\ 1.$$

So, if you want to know how many different ways there are to choose k things from a group of n, the answer is the kth element in the nth row in Pascal's triangle. Here's an example. How many different ways are there to choose 5 paintings from 7 that are on display? $n = 7$, $k = 5$ and just go to the fifth number in the seventh row in Pascal's triangle to get the answer: 21 (note that there is a column 0 and a row 0).

Mathematicians use the following notation to symbolize "choose k elements from a set of n":

$$\binom{n}{k},$$

which is read "n choose k". There is also a specific formula to calculate this: using the *factorial*, a mathematical operator denoted by an exclamation mark. The factorial of a (whole) number is the product of all the whole numbers up to it, so:

> 0 factorial, denoted: 0! equals 1 (by definition)
> $1! = 1$
> $2! = 1 \times 2 = 2$
> $3! = 1 \times 2 \times 3 = 6$
> $4! = 1 \times 2 \times 3 \times 4 = 24$

and so on. Now you have all the ingredients to understand the explicit formula for the binomial coefficients:

$$\binom{n}{k} = \frac{n!}{k!(n-k)!} \qquad (5.4)$$

So, plugging in $n = 7$ and $k = 5$ for the painting puzzle, we get,

$$\binom{7}{5} = \frac{7!}{5!2!} = \frac{1 \times 2 \times 3 \times 4 \times 5 \times 6 \times 7}{(1 \times 2 \times 3 \times 4 \times 5) \times (1 \times 2)} = \frac{5040}{240} = 21$$

which is just the same as going to the 5th number in the 7th row in Pascal's triangle. Indeed, many times, instead of using numbers, all of the entries in Pascal's triangle are written using the "n choose k" notation:

$$\binom{0}{0}$$

$$\binom{1}{0}\binom{1}{1}$$

$$\binom{2}{0}\binom{2}{1}\binom{2}{2}$$

$$\binom{3}{0}\binom{3}{1}\binom{3}{2}\binom{3}{3}.$$

We've strayed quite a bit from the number pyramid puzzle. Let's take a more careful look at it. It certainly resembles Pascal's triangle, with one important difference. Instead of regular addition, two numbers are added in a different way — taking only the units digit once the summation is over. A more mathematical way of stating this is: the numbers are added *modulo* 10, a concept introduced by the famous German mathematician, Carl Friedrich Gauss, who first wrote about modular arithmetic in his book, *Disquisitiones Arithmeticae*, published in 1801. A scrutinizing study by Maarten Bullynck lays out the foundations of this research in earlier works on the math of remainders in division problems by many prominent mathematicians including Claude Gaspard Bachet de Méziriac, Leonhard Euler, Carl Friedrich Hindenburg and many others. Gauss was a prominent mathematician and physicist in the 18th and 19th centuries, who made important discoveries in a huge number of fields, from magnetism, astronomy and geodesics to number theory and differential geometry. However, he is mostly known to kids today as the kid who amazed his teacher at the age of seven, by summing the numbers 1 to 100 in his head. Gauss's method was based on the fact that $1 + 100 = 2 + 99 = 3 + 98 = 4 + 97 = \ldots = 101$, so to get the sum of the numbers from 1 to 100, all you need to do is multiply 101 by the number of pairs, which is 50, giving a correct answer of 5050.

Modular arithmetic is sometimes called *clock arithmetic* because you can visualize it on the face of a clock. $10 + 3 = 13$, of course. But on the face of a 12-hour clock, adding 3 hours to 10, gives us 1. Similarly, adding 2 to 12 gives us 2, adding 6 to 9 gives us 3 and adding 11 to 5 gives us 4. This kind of adding is called *adding modulo* 12 so we can write mathematically:

$$(10 + 3) \bmod 12 = 13 \bmod 12 = 1$$
$$(12 + 2) \bmod 12 = 14 \bmod 12 = 2$$
$$(9 + 6) \bmod 12 = 15 \bmod 12 = 3$$
$$(5 + 11) \bmod 12 = 16 \bmod 12 = 4.$$

Take another look at the numbers, this time just the result of the summation modulo 12 (the third column in the expressions above). You'll realize that what the modulo operation is really doing is taking the remainder that is left after dividing the number by 12:

$$13 \bmod 12 = \text{the remainder of } 13 \div 12 \text{ which is } 1$$
$$14 \bmod 12 = \text{the remainder of } 14 \div 12 \text{ which is } 2$$
$$15 \bmod 12 = \text{the remainder of } 15 \div 12 \text{ which is } 3$$
$$16 \bmod 12 = \text{the remainder of } 16 \div 12 \text{ which is } 4.$$

We can imagine a clock with 10 numbers on its face, instead of 12. In this case:

$$(10 + 3) \bmod 10 = 13 \bmod 10 = 3$$
$$(7 + 2) \bmod 10 = 9 \bmod 10 = 9$$
$$(9 + 7) \bmod 10 = 16 \bmod 10 = 6$$
$$(5 + 7) \bmod 10 = 12 \bmod 10 = 2.$$

This is exactly the kind of summation we were using in the puzzle, so taking the units digit after summing two numbers is equivalent to taking the remainder of the sum when divided by 10 — in other words taking the sum *modulo* 10.

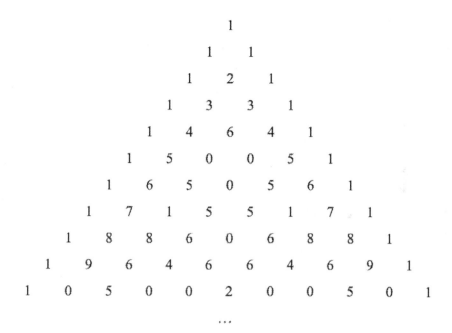

Fig. 5.12   Pascal's triangle modulo 10

One thing often gets me confused — the zero. Mathematically, when defining rigorously the modulus as the remainder after division, the results will always be a number between 0 and the modulus minus 1. So, 10 modulo 10 is equal to 0. However, when doing *clock arithmetic* in real life, the modulus itself is the "zero". That's why 6 + 6 on a clock is 12 and not 0.

It's interesting to take a look at Pascal's triangle modulo 10, shown in Fig. 5.12. This gives us the real solution to the whole puzzle! The numbers in the modulo-10 Pascal's triangle are the coefficients that we need to multiply by the numbers in the given line to get the correct apex number. Recall that the given line in the puzzle had 11 numbers. Now look at the eleventh line in the modulo-10 Pascal triangle in Fig. 5.12. This tells us that to get the right answer we need to take the first numbers and add to it five times the third number, twice the sixth number, five times the ninth number and once the eleventh number to get the correct apex number. We could have started the question with a line of five numbers, but then, to calculate the apex number we would need to take the sum of the first number, add to it four times the second number, six times the third number, four times the fourth number and once the last number in the row, a hard feat to do in your head, quickly. An easier option would be to pose the puzzles with just six numbers. Then the quick answer is obtained by summing the first and last numbers together with five times the second and fifth numbers. If the line was: 12 4 5 82 8 13, the answer would be: $(12 + \mathbf{5} \times 4 + \mathbf{5} \times 8 + 13)$ mod 10 which gives the apex number: 5. The key to the puzzle is to choose the best row to start with, the one that will make the prediction easy, in this case, either the sixth or eleventh row.

Generalizations of the Puzzle

Historically, this puzzle is a generalization of a magic trick using playing cards and based on a different modulus — in fact two versions, one based on modulo-9 and the other modulo-3. The trick is well described by a number of people including Martin Gardner, who was, as far as I know, the first to write about it in his popular book, *Mathematical Carnival*. Card magician and mathematician, Colm Mulcahy, has also written about the trick, and so have Ehrhard Behrends and Steve Humble, who published a mathematical analysis of the modulo-3 version in the *Mathematical Intelligencer*. The modulo-3 version can be done in various ways, but a particularly nice way is to involve a large audience — and do it as a magic trick.

First, let's find out which rows are those that make the prediction easy. To do this we'll create Pascal's triangle modulo-3 using a spreadsheet. To make things easier with the zero, we'll use the numbers: 0, 1 and 2, and the relationships:

$$(0 + 0) \bmod 3 = 0 \bmod 3 = 0$$
$$(0 + 1) \bmod 3 = 1 \bmod 3 = 1$$
$$(0 + 2) \bmod 3 = 2 \bmod 3 = 2$$
$$(1 + 1) \bmod 3 = 2 \bmod 3 = 2$$
$$(1 + 2) \bmod 3 = 3 \bmod 3 = 0$$
$$(2 + 2) \bmod 3 = 4 \bmod 3 = 1.$$

Since I wanted to make an 82-row triangle, and don't want to ruin your eyesight, I've color-coded the numbers:

0 is colored white

1 is colored gray

2 is colored black

You can see the result in Fig. 5.13.

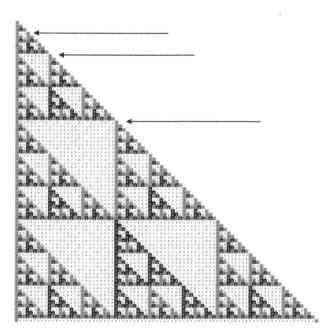

Fig. 5.13 Pascal's triangle modulo 3

Before we go on, stand back a bit and admire the beautiful artwork of the triangle! This kind of structure is known as a fractal. Let's search the fractal for those lines with "lots of zeros", colored white. I've saved you the trouble — it's the ones with the arrows — lines 4, 10, 28 and the last line, 82 (no arrow there for lack of space). What's even nicer is that in these four rows, the only numbers that are not zero are the first and last, and the numbers are 1. So, for this version of the puzzle, if you start with a line with either 4, 10, 28 or 82 numbers, the correct prediction for the apex number is just the sum of the first and last numbers in the row.

Now, for the large audience magic trick! First, you need to arrange the people in a triangle, so that one person stands at the back, two in front of him, three in the third row from the back and so on, until there are 4, 10, 28 or 82 people in the front.

Each person in the first row has a card on which he writes one of three numbers, 0, 1 or 2, and holds it up. The magician writes down a prediction which is, of course, the modulo-3 sum of the cards in the hands of the two people at the ends of the row. The rest of the procedure is just building the pyramid. Each person in the second row now does the math. He looks at the cards of the two people directly in front of him.

- If both cards are 0, he writes on his card 0.
- If both cards are 1, he writes on his card 2.
- If both cards are 2, he writes on his card 1.
- If one card is 0 and the other 1, he writes on his card 1.
- If one card is 0 and the other 2, he writes on his card 2.
- If one card is 1 and the other 2, he writes on his card 0.

This procedure carries on, row after row until the apex person reveals his card, which, naturally, is the same as the prediction. If you want to avoid numbers, you can use colors instead and derive the equivalent "color sums" instead of the six rules. If you want to make a more impressive effect, when you make your prediction you can say that the odds of you getting it right is 1:3, which would be true if you were randomly guessing the answer.

Ehrhard Behrends and Steve Humble realized that the "easy rows" — those that had just two 1's on each end with the rest of the numbers all 0's — occurred at interesting intervals: 4, 10, 28 or 82. These numbers are all one number larger than the sequence of powers of 3: 1, 3, 9, 27, 81. So now we can predict that the next "easy row" will be row number $3^5 + 1 = 244$. Moreover, *every* single row pattern in Fig. 5.13 repeats itself every $(3^n - k)$ row, where n and k are whole numbers. For example the pattern in row 7 repeats itself in rows 25, 79 and 241. Each pattern is replicated three times longer, which creates the beautiful fractal picture. But what are fractals?

Fractals are patterns that are infinitely complex, and self-similar at different scales. Take a step back and look at the triangle in Fig. 5.13 as a picture, rather than a number pyramid. It looks like the same complex pattern, over and over again, at different magnitudes. The pattern can be described in the following manner: take a black-colored right-triangle, divide it into nine, equal-sized right-triangles and remove the three triangles that are upside-down. Do exactly the same for each black-colored triangle. Continue until infinity.

Fractal analysis is a "hot" and extremely important branch of math, in particular, because generating infinitely complex, self-similar patterns is a great way to model real-life objects. The person who first recognized this and is acknowledged as the "founder" of fractals was the 20th century American–French–Polish mathematician, Benoit Mandelbrot, who realized that a couple of centuries of work on self-similar structures, done by prominent mathematicians such as Georg Cantor, Helge von Koch and Wacław Sierpiński, were in fact all connected in this one concept, that he termed *fractal*. The following figures show some famous fractals. The Sierpiński fractals are all generated in a similar way to the number pyramid fractal. Take a colored shape, divide it into sub-shapes and remove some of them. Cantor's fractal, known as the

Cantor set, is one of the simplest, yet interesting fractals you can imagine. It is generated by taking a line segment, mentally dividing the segment into three and removing the middle segment. That's all — just continue to infinity. Von Koch's snowflake is generated as follows:

- Begin with an equilateral triangle.
- Mentally divide every line segment into three segments of equal length.
- Replace the middle segment with an equilateral triangle, each side of which is the same size as the middle segment, and remove its base.
- Continue to infinity with every line segment in the shape.

All these fractals, and many others, have 3-dimensional analogs.

Mandelbrot laid down the mathematical foundations for generating fractals through a set of rules, just like those we used for the number pyramid fractal, or those used to generate the fractals in Figs. 5.14–5.17. He also found ways to mathematically describe some of their features. A key concept that he invented is the *fractal dimension*.

Fig. 5.14   Five iterations of Sierpiński's triangle

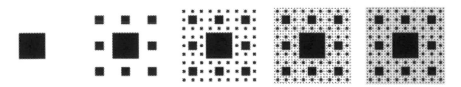

Fig. 5.15   Five iterations of Sierpiński's carpet

Fig. 5.16   Seven iterations of Cantor's set

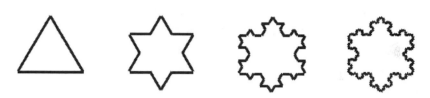

Fig. 5.17   Four iterations of Koch's snowflake

In classical geometry, we are aware of the integer dimensions: the one-dimensional line, the two-dimensional square, the three-dimensional cube, etc. These are known as topological dimensions. In contrast the fractal dimension of an object is not confined to whole numbers. The fractal dimension of the Sierpiński triangle is 1.58..., that of the Sierpiński carpet is 1.89..., and the Koch snowflake has a fractal dimension of 1.26... . Calculating fractal dimensions is hard, and to make it worse, mathematicians have suggested many, different definitions, not all of which converge to the same value! For simple fractals, though, like those above, there is a method which, with some practice, can be mastered. Notice that the outline of each fractal stays the same size throughout the iterations. However, it is made up of smaller versions or units of itself, magnified. Indeed, this is the whole essence of fractals! So, for example, if you take any one of the three triangles in the third iteration of the Sierpiński triangle, and "blow it up", i.e. magnify it by a factor of 2, you'll get exactly the same shape as the second iteration. The fractal dimension is calculated by dividing the logarithm of the number of self-similar units of the fractal by the logarithm of the magnitude factor. For those of you unfamiliar with logarithms, all you really need to know is that it's just a mathematical operator, like squaring or cubing or taking the square root, that can be found on any calculator. It is the opposite of the exponential function (formally under any base system), i.e. if $10^2 = 100$ and $10^3 = 1{,}000$, the logarithms of 100 and 1,000 are 2 and 3 respectively.

Going back to the fractal dimension calculation, and using N to mean the number of self-similar units and M the magnitude factor between two consecutive iterations, the fractal dimension is $\log(N) \div \log(M)$. For the Sierpiński triangle, $\log(3) \div \log(2) = 1.58...$ This works at any scale, or any stage of the iteration, and indeed if we consider the fourth iteration of Sierpiński's triangle, we get $\log(9) \div \log(4) = \log(3) \div \log(2) = 1.58...$.

Let's calculate the fractal dimension for the Sierpiński carpet. The number of self-similar units in the third iteration is 8 and the magnitude factor is 3, because if you "blow-up" one of the 8 self-similar units 3 times, you get the fractal shape from the previous step. The fractal dimension is therefore $\log(8) \div \log(3) = 1.89...$. For the Cantor set, the fractal dimension is $\log(2) \div \log(3) = 0.63...$.

What do these numbers all mean? There is a nice intuitive explanation. The topological dimension of a line segment is 1. Now, Cantor's set, after an infinity of iterations, is made up of an infinite number of line segments, yet they will never, ever, make up a full line segment. The infinite Cantor set will look more like a row of unevenly distributed dust particles! So perhaps it would be more "accurate" in a way, to say that it is something between a 0-dimensional point and a 1-dimensional line segment, which indeed it is. Similarly, Sierpiński's carpet takes up more space than a line segment, but less than a square, so its dimension should be something between 1 and 2. The underlying philosophical justification of fractal dimensions is the fact that fractals are complex, infinitely.

One of the things that got Mandelbrot interested in fractals was his urge to be able to create a realistic model of Nature on a computer, a feat that in 1979 seemed impossible. He argued that classic geometry can't describe real-world objects like broccoli, or

Fig. 5.18   Computer-generated fern (left) vs. real-life fern (right)

a mountain landscape, or the coast of Britain, and looked for other alternatives, stumbling into fractal geometry. He discovered that, not only can fractal geometry model these things really well, as you can see from the computer-generated fern compared with a real fern in Fig. 5.18, it works in the opposite direction as well!

Broccoli, mountain landscapes, the coast of Britain, and many infinitely complex, real-life objects can be described using fractals. Many computer-generated fractals are beautiful, and useful! The movie industry uses fractals to generate real-life looking scenes, like the mountains in the James Bond movies, or the moon in Apollo 13, and apologies for ruining it for those who thought they were real.

We've strayed quite a long way from the number pyramid fractals. To come full circle, Figs. 5.19 and 5.20 present the fractals for the modulo-10 number pyramid puzzle, the one we started from, and for the particularly interesting modulo-9 puzzle. For the latter, you can start with 10 numbers in the original line, and the apex number can be calculated by taking the sum of the first and last numbers, together with thrice the fourth and seventh as you can readily see from the least unshaded lines in the fractal. These fractals have a more complicated, and mathematically interesting structure than the modulo-3 case, with some protruding colored lines that look like they were randomly scattered throughout the fractal — a feast for a researcher. Further generalization to other moduli and larger initial numbers is now easy to do, by generating the appropriate fractals.

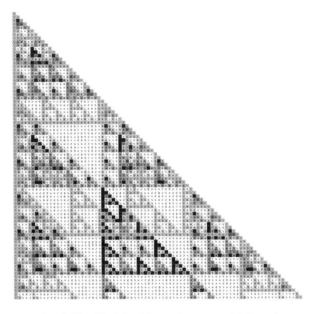

Fig. 5.19   Modulo-10 number pyramid fractal

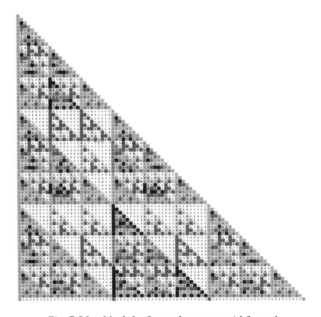

Fig. 5.20   Modulo-9 number pyramid fractal

Recap

In this chapter we started off with a number pyramid puzzle that can also be performed as a magic trick. We saw the connection between number pyramids and Pascal's triangle, and along the way we learnt about some of the interesting number sequences hidden in the triangle and the significance of Pascal's triangle in combinatorics and binomial expansion. We discussed some of the math of different moduli and the beautiful fractals that moduli number pyramids form.

Here is a brief summary of some of the concepts we encountered along the way.

- $\binom{n}{k}$ — the number of ways to choose k elements from a set of n.

- *Apex number* — the number at the top of a number pyramid.

- *Binomial coefficient* — the numbers that multiply the powers of x and y in the binomial theorem.

- *Binomial theorem* — the algebraic expansion of $(x + y)^n$.

- *Combinatorics* — the area of math that has to do with counting things.

- *Fibonacci sequence* — the series of whole numbers: 0, 1, 1, 2, 3, 5, 8, 13, etc. where every number in the sequence is the sum of the two previous numbers.

- *Figurate numbers* — numbers that, when replaced by points or dots, can be arranged in a geometric arrangement.

- *Fractal dimension* — a measure, usually not a whole number, of how a fractal fills space as it scales.

- *Fractals* — patterns that are infinitely complex, and self-similar at different scales.

- *Modular arithmetic, clock arithmetic* — the area of math that involves a closed, cyclic set of consecutive whole numbers, as appears on the face of a clock. Also, the remainder after division by one of the numbers in the set.

- *Natural numbers* — the positive integers, used for counting and ordering.

- *Number pyramid* — a pyramid of numbers, where every two adjacent numbers are summed modulo a given number, and the result is horizontally positioned in between them in the row directly above.

- *Pascal's triangle* — a number pyramid, with the apex number 1, and every two adjacent numbers are summed, with the result positioned horizontally in between them in the row directly below.

- *Repunit* — a series of a repetition of ones.

- *Simplex numbers* — triangle figurate numbers in different dimensions.

- *Square numbers* — square figurate numbers, also the product of an integer and itself.

- *Tetrahedral numbers* — tetrahedron figurate numbers, also the cumulative sum of the consecutive triangular numbers.

- *Triangular numbers* — equilateral triangle figurate numbers, also half the product of two consecutive whole numbers.

- *Whole numbers* — the natural numbers and zero.

Challenge Yourself!

1. Mentally predict the apex number of a modulo-10 number pyramid if the eleven initial numbers are

$$3 \ 18 \ 8 \ 0 \ 9 \ 18 \ 50 \ 17 \ 2 \ 39 \ 6.$$

2. Mentally predict the apex number of a modulo-9 number pyramid if the ten initial numbers are

   1 2 0 3 4 2 12 8 19 40.

3. What would you consider to be the first six pentagonal numbers?

4. Use the binomial theorem to calculate 13^5 easily, without using a calculator.

5. Use Pascal's triangle to find the first five 5-simplex numbers.

6. Use Pascal's triangle to find out how many handshakes there are in a room of ten people, if everyone shakes hands once with everyone else.

7. Draw a fractal tree using the following steps. Draw the letter Y. On each of the two branches of the letter Y draw another letter Y. Continue forever!

8. What is the fractal dimension of the Menger sponge? The first four iterations of the sponge can be seen in Fig. 5.21.

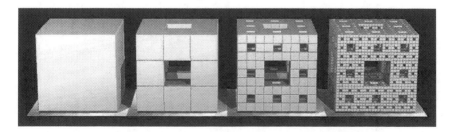

Fig. 5.21   First four iterations of the Menger sponge

9. Use a spreadsheet program to create a modulo-7 number pyramid fractal.

Solutions

1. The apex number of a modulo-10 number pyramid where the eleven initial numbers are

    3 18 8 0 9 18 50 17 2 39 6

    is the units digit of $3 + 6 + 2 \times 18 + 0 + 0$, or, put in another way, $(3 + 6 + 2 \times 18 + 0 + 0) \bmod 10 = 45 \bmod 10 = 5$.

2. The apex number of a modulo-9 number pyramid where the ten initial numbers are

    1 2 0 3 4 2 12 8 19 40

    is the units digit of $1 + 3 \times 3 + 3 \times 12 + 40$, that is $(1 + 9 + 36 + 40) \bmod 9 = 5$.

3. The first six pentagonal numbers are 1, 5, 12, 22, 35, 51.

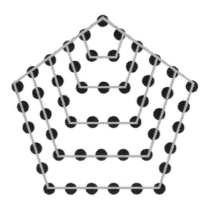

Fig. 5.22   The pentagonal number 51

4. Rewrite 13 as (10 + 3) and use the fifth row of the Pascal triangle:

$(10 + 3)^5$
$$= 10^5 + 5 \times 10^4 \times 3 + 10 \times 10^3 \times 3^2 + 10 \times 3^3 \times 10^2 + 5 \times 3^4 \times 10 + 3^5$$
$$= 10^5 + 15 \times 10^4 + 9 \times 10^4 + 27 \times 10^3 + 5 \times 81 \times 10 + 243$$
$$= 10^5 + 24 \times 10^4 + 27 \times 10^3 + 405 \times 10 + 243$$
$$= 371293$$

5. The first five 5-simplex numbers appear along the sixth diagonal and they are 1, 6, 21, 56, 126.

6. There are $\binom{10}{2} = 45$ handshakes.

7.

Fig. 5.23   A simple fractal tree

8. The fractal dimension of the Menger sponge is

$$\log(20) \div \log(3) = 2.72...$$

9.

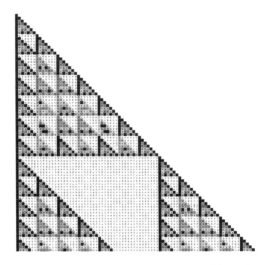

Fig. 5.24   Modulo-7 number pyramid fractal

Bibliography and Further Reading

Behrends, E., & Humble, S. (2013). Triangle Mysteries, *The Mathematical Intelligencer*, **35**(2), pp. 10–15.

Bullynck, M. (2009). Modular arithmetic before CF Gauss: Systematizations and discussions on remainder problems in 18th-century Germany, *Historia Mathematica*, **36**(1), pp. 48–72.

Elran, Y. (2014), The generalized apex magic trick, *Recreational Mathematics Magazine*, **1**, pp. 53–58.

Gardner, M. (1989). Pascal's triangle, in *Mathematical Carnival*, Mathematical Association of America, Washington DC.

Jones, M. A., Mitchell, L., & Shelton B. (2015). Fractals & mysterious triangles, *Math Horizons*, **23**(1), pp. 22–25.

Mitchell, L., Jones, M. A., & Shelton, B. (2016). Abelian and non-Abelian triangle mysteries, *The American Mathematical Monthly*, **123**(8), pp. 808–813.

Mulcahy, C. (2012). All or Nothing Trickle Treat, *Card Colm*. http://cardcolm-maa.blogspot.co.il/2012/10/all-or-nothing-trickle-treat.html

Neville, G. (2005). The pyramid collection, *Australian Mathematics Teacher*, **61**(1), pp. 9–13.

O'Connor, J. J., & Robertson, E. F. (1996). Blaise Pascal, *MacTutor History of Mathematics Archive*. https://mathshistory.st-andrews.ac.uk/Biographies/Pascal/

chapter 6

Puzzles from the Pandemic: Math and COVID-19

Introduction

This chapter wasn't in the original sketch of the book. But 2020 came, and with it — a bombshell — the COVID-19 coronavirus pandemic, and we just couldn't ignore it. From China to the United States, no country was safe. Humanity had not faced such a challenge since the Spanish flu, one hundred years ago. The pandemic took my own country, Israel, by surprise in March that year. World leaders and politicians had to make decisions — fast. Many dilemmas arose, but it quickly became apparent that math had at least some of the answers to the questions faced in real life. Israeli scientists and mathematicians from the Weizmann Institute of Science and elsewhere, formed an advisory committee to the government. Led by physicist Eli Waxman, the committee helped the government make decisions based on real data, and math. At one point, the committee approached the Davidson Institute of Science Education, where I work, and asked us to help them communicate with the public with regards to decisions based on facts, rather than "fake news". Our online articles became very popular and were soon adopted by the press. They were hailed by the committee as being instrumental in combatting many misconceptions. I was privileged to work hand-in-hand with such leading Israeli scientists and mathematicians.

The real-life puzzles that faced the mathematicians were many. What is the correct model for the pandemic? Is it worthwhile to aim for "herd immunity"? And then came the lockdowns, and the masks, and the social distancing, and with them, more puzzles. For example, how many people can fit in a room of a given size such that they keep a distance of 2 meters from each other? So many puzzles, and a heaven for mathematicians! I've chosen one of the puzzles, in fact, the very first puzzle that COVID-19 presented, as the "show" puzzle for this section since, funnily enough, it mimics a very old riddle about grains of wheat on a chessboard! But afterwards, we'll dive into some of the other puzzles that we dealt with, and journey into mathematical modeling, big data and some other cool areas of math.

The Puzzle

First Puzzle: It's March 2020. The Prime Minister of Israel, Benjamin Netanyahu, is holding a dramatic press conference. "The situation is serious", he begins. "There is a contagious and lethal virus at large — COVID-19. It progresses like a geometric series. One person infects another. Each of them infects one more person. The four contagious people infect eight, eight infect sixteen, sixteen infect thirty-two. If we don't do anything at all, the numbers will be outrageous."

There are 9,000,000 people in Israel. Assuming Netanyahu is right, and assuming that an infection occurs every hour, after how many hours will the whole population of Israel be infected? (Calculate to the closest hour where this number of people will definitely hold.)

Second Puzzle: Netanyahu is wrong — because when about half the population is infected, an effect known as "herd immunity" will kick in and the pandemic will be over.

How many hours will it take before half the population is infected? Assuming that 2.5% of the population is hospitalized, how many people will be hospitalized if herd immunity is reached and 4,500,000 are infected. Assuming 30% of the hospitalized unfortunately dies, how many people will that be?

Where to Start?

This puzzle is based on a *geometric series*, a series of numbers with a constant ratio between every two consecutive numbers. In our puzzle, the first numbers of the series are given: 1, 2, 4, 8, ... so it is easy to calculate the constant ratio and extrapolate to get the correct answer.

Solving the Puzzle

The constant ratio is $8 \div 4 = 4 \div 2 = 2 \div 1 = 2$. Therefore, the full series up to over 9,000,000 is:

1, 2, 4, 8, 16, 32, 64, 128, 256, 512, 1024, 2048, 4096, 8192, 16384, 32768, 65536, 131072, 262144, 524288, 1048576, 2097152, 4194304, 8388608, 16777216.

Counting the number of elements (numbers) in the series gives us 25 hours, so it will take exactly 24 hours to infect all the population. You might be surprised it's 24, not 25, but this is just one of those things math is notorious for — you have to be meticulous, otherwise you won't get a true answer. The first person who gets infected doesn't count, since an hour hasn't passed for him to get infected, so after one hour the second person gets infected, after two hours four people are infected and after one whole day, the whole population will have been infected. We're assuming here that we're not chopping up hours into minutes, so the number we need to take as "the whole population" is the first number in the series larger than the exact population (9,000,000) which is 16,777,216. This can be justified since we are not told *how* the people are infected, i.e. the frequency of their infective interactions, so it might be that most of them are infected very much near the end of the hour, i.e. the 9 millionth person becomes infected at the last minute along with 16,777,216 – 9,000,000 others ...

Surprisingly, for half the population to become infected, it takes only one hour shorter, i.e. 23 hours, since after 23 hours 8,388,608 people will be infected, and 8,388,608 is the first number in the series larger than 4,500,000, which is exactly half the population.

Notice that this geometric series can also be written as the consecutive powers of 2, since $2^0 = 1$, $2^1 = 2$, $2^2 = 4$, $2^3 = 8$, ..., $2^{24} = 16,777,216$. This suggests a formula to find the number of people infected after any given hour: 2^h, where h is the number of hours. To do the opposite, i.e. find the number of hours it will take to reach a given number, you have to take the *base-2 logarithm* of the number, which you can do using a scientific calculator (base-10 logarithms were discussed in Chapter 5. There is a definition of base-2 logarithm in the Recap section of this chapter).

Calculating the percentage of people who will be hospitalized if 4,500,000 become infected, and how many of those will unfortunately die, is straightforward:

$$2.5\% \text{ of } 4,500,000 = 2.5 \times 4,500,000 \div 100 = 112,500$$
$$30\% \text{ of } 112,500 = 30 \times 112,500 \div 100 = 33,750.$$

For a small country like Israel, these numbers make herd immunity unfeasible. The health system cannot cope with anything close to 100,000 people; 2,000 COVID-19 hospitalizations is the absolute upper limit for the whole of Israel. Above that, there are just not enough hospital beds, doctors, nurses, breathing aid devices, medicine and other medical necessities. Moreover, the huge influx of over 100,000 severe cases of COVID-19 that will have to go without medical care, means that the percentage of people who will unfortunately die will be much more than 30% and far more than the projected estimate of 33,750, itself a huge number for a small country (compare for example with the mortality rate of influenza and pneumonia of about 1,250 people annually).

The History of the Puzzle and Related Topics

One might think that there can't possibly be a history of a real-life puzzle. Yet, interestingly enough, the puzzle is ancient. Sometimes known as the "wheat and chessboard" or "rice and chessboard" puzzle, it has its origins in Indian or Arabian math. Here's one version.

When an Indian ruler asked the mystical, 5th century, Brahmin Sissa, inventor of chess, what his wages are, he answered, "Bring me a chessboard!" He then said that his fee is the sum of the grains of wheat placed on the chessboard in the following manner: one grain on the first square, two on the second, four on the third, eight on the fourth, and so on, doubling the number of grains on each successive square. The ruler laughed at such a meager price. The last laugh was on him. The wages were astronomical, 18,446,744,073,709,551,615 to be precise. This puzzle is constantly regurgitated in different forms in the puzzle world.

A puzzle similar to the second part of our COVID-19 puzzle is the lily pond puzzle. A lily in a pond doubles its size every day until, after 15 days, it fills the whole pond. On what day did the lily fill half the pond? The non-intuitive answer, of course, is 14. People are always taken by surprise when a geometric series gets into the large numbers. The ruler in the chess puzzle was taken by surprise — and we were too when we experienced the rapid infection rate of COVID-19.

Fig. 6.1   Sissa ibn Dahir (artist's impression)

The concept of geometric series began much earlier than the puzzle, dating far back to Egyptian and Greek math, hundreds of years BCE. One case of a geometric series is recorded in a famous papyrus, the Rhind mathematical papyrus, named after the Scottish archaeologist and antiquarian Alexander Henry Rhind who acquired it in 1863. The papyrus, now on display at the British Museum in London, was written by a scribe named Ahmes in the 16th century BCE. It is one of the most valuable historical math artifacts, covering many topics in ancient Egyptian math, like arithmetic, algebra and geometry. The papyrus is full of problems and their solutions, providing math historians with a great perspective of math in ancient times.

Problem 79 in the papyrus is the following (rewritten using modern terminology). Someone owns 7 houses, 49 cats, 343 mice, 2401 sheaves of wheat and 16807 measures of grain. How many items did he own, altogether? There are a number of ways to calculate the answer. At first, it looks just like the usual arithmetic sum:

$$7 + 49 + 343 + 2401 + 16807 = 19607,$$

and indeed this is Ahmes's first solution to the problem. However, notice that the numbers are all powers of 7:

$$7^1 = 7, 7^2 = 49, 7^3 = 343, 7^4 = 2401, 7^5 = 16807.$$

Also notice that the sum of a geometric series of length n is equal to the sum of the first $n - 1$ numbers plus 1, multiplied by the constant ratio. For example, the sum of the first three numbers in the series is equal to the sum of the first two numbers plus 1, times 7:

$$(7 + 49 + 1) \times 7 = 57 \times 7 = 399.$$

So the sum of all the numbers in Ahmes's puzzle is equal to

$$(7 + 49 + 343 + 2401 + 1) \times 7 = 19607,$$

the correct answer. Indeed, Ahmes used this method as a second way of solving the problem, hinting to the fact that he knew about geometric sequences.

Fig. 6.2 A fragment of the Rhind Papyrus

We could now dive into Ahmes's papyrus and Egyptian math, which is absolutely fascinating, but I don't want to stray too far from pandemic math, so we'll take a huge leap forward to Euclid, mathematics' bestselling author to date — whom we got acquainted with in the first chapter. Euclid was a 2nd century BCE Greek mathematician, best known for his 13-book work called *The Elements*, in which he laid down the foundations of math, which was taught for 2000 years. Although the Egyptians, Babylonians and Greeks all used the geometric series, Euclid's work is the first known rigorous treatment of geometric series which he writes about in books 8 and 9 in *The Elements*. It was from the 14th century CE onwards, that more detailed analysis and further expressions and uses were given for geometric series, mainly by the French mathematician, Nicole Oresme, and the British mathematicians Richard Swineshead (in the 14th century), and John Wallis and Isaac Newton (both in the 17th century).

The geometric series is *not* the correct way to calculate infection rates. The reason is that the ratio between the number of people infected with time is not constant — it decreases. This is because, as more and more people are infected, they become immune, at least for a period of time, so the number of interactions a virus has with non-immune people becomes lower and lower, until herd immunity is reached. When about half, or perhaps a bit more, of the population is immune, the people who *are* still infected, will most probably never meet anyone who is not immune, and the virus will just "die alone".

Government-imposed lockdowns and social isolation work the same way. They minimize interactions, so the virus does not have a chance to spread. In an ideal world, and some places like New Zealand, this is enough to eradicate the virus completely, at least for a while. However, in most cases, when the limitations are removed before the virus has been defeated, people will not comply 100% with the rules and infection rates will start soaring again in second and third waves.

Can math *truly* model the spread of infectious viruses? They can at least come close, and we have a Belgian mathematician, Pierre François Verhulst, to thank for that. Verhulst was born in Brussels in 1804, and was educated at the Athenaeum, one of the best schools of the time. He excelled in math, science and poetry, and was privileged to have another great Flemish mathematician, Adolphe Quetelet, as his teacher and mentor. He had a strong social conscience, and had the proclivity to get into trouble for professing unorthodox views, to the point where he was expelled from Italy in 1830 — he was at the time on a trip to Rome — for trying to persuade the Pope to adopt a constitution in the Papal States. Verhulst held academic positions at the Belgian Military Academy, after enlisting in the newly-assembled independent Belgian army, and for a period of time also at the Université Libre of Brussels. Between 1838 and 1847, Verhulst published three important papers that became the foundation of population dynamics — a mathematical description of how real-life systems behave.

Fig. 6.3   Pierre François Verhulst

In his first paper, Verhulst explained the rationale for his work, which resonates with the reason why the geometric series is just not enough. He was looking at the way populations grow, and argued that they will never grow indefinitely since there are limiting factors that have to be considered, like shortage of food, or insufficient living conditions, etc. He wrote, "We know that the famous Malthus showed the principle that the human population tends to grow in a geometric progression so as to double after a certain period of time, for example every twenty-five years. This proposition is beyond dispute if abstraction is made of the increasing difficulty to find food ... The virtual increase of the population is therefore limited by the size and the fertility of the country. As a result, the population gets closer and closer to a steady state ...". Later on, in a paper published in 1845, he again reasoned against using the geometric progression, "We shall not insist on the hypothesis of geometric progression, given that it can hold only in very special circumstances; for example, when a fertile territory of almost unlimited size happens to be inhabited by people with an advanced civilization, as was the case for the first American colonies."

Verhulst proposed an alternative, the *logistic function*, the solution to what's known as the logistic equation. The logistic equation is a *differential equation*, a genre of equations that was invented not long before Verhulst's time, by both Isaac Newton and Gottfried Wilhelm Leibniz, independently. Differential equations describe the rate of change, usually how something changes with time. In the case of the logistic function, how populations of people, animals, plants, viruses, etc. change with time, considering factors that contribute to population growth, and those that cause population decline. Of course, the parameters that describe these different factors are specific to each case, so we'll use different parameters depending on whether we're looking at the change in COVID-19 infection (i.e. virus population), or the rate of change of the population of India, or that of Iceland.

Throughout the years, many different logistic functions have been suggested. In 1959, British botanist Francis John Richards, a member of the Royal Society and a professor at Imperial College, London, suggested a *generalized logistic function,* a "master equation" from which logistic models can easily be derived for nearly every imaginable case. It is from this equation that COVID-19 scientists derive their models for the pandemic. Rami Band, a friend and mathematician at the "Technion" – the Israel Institute of Technology, pointed out to me that the specific model that is used for pandemics (and COVID in particular) is called SIR. It stands for Susceptibles, Infected and Recovered. This model is relatively basic and goes back to 1927. These days this model is extended in various ways, but essentially all extensions are related to this basic SIR.

Logistic functions all have the same looking graph when we plot the population as a function of time — the sigmoid — an S-shaped curve, like the one shown in Fig. 6.4. If you look at the sigmoid, you'll see that initially, the population grows very fast, but there's a point where "the tables turn". In the middle of the S, the curve

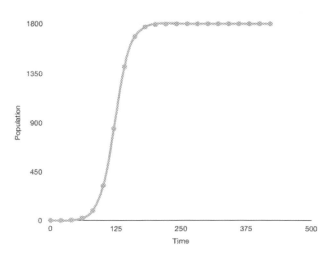

Fig. 6.4　A sigmoid curve

takes a turn and starts leveling out, due to the increasing impact of the second term in the equation, finally reaching a plateau, where the population doesn't grow anymore. The S curve can be steep or stretched. This depends on what's known as the *growth rate*, a parameter used in the equation. There is one major problem that hinders these models. People cannot really be modeled! So, although we can interpret the data as it comes in, we cannot predict when the curve will flatten, when we'll reach the plateau of the sigmoid, or even when this plateau will occur, i.e. in the case of COVID-19 or other pandemics, what the final number of infected people will be. In fact, we can, ourselves, influence this by changing our behavior. Maintaining social distance, washing our hands and wearing masks, all reduce the rate of infection, and help to "flatten the curve".

Generalizations of the Puzzle

A generalization of the puzzle, indeed something that researchers really did, was to see how well different logistic models actually model COVID-19. Efim Pelinovsky, chief scientist at the Institute of Applied Physics at the Russian Academy of Sciences, along with a group of Russian scientists, did exactly this, and found remarkably that in some cases, even Verhulst's crude logistic model quite accurately reproduced what happened in some countries during the first two months, the "first wave" of COVID-19. We'll do the same. We'll take Turkey as an example and we'll use Verhulst's original function. It is what's known as an *ordinary differential equation*, an equation that describes the rate of change of a population, where the population can be the number of people/plants/animals, etc. or in our case, the total number of people who are COVID-19 positive. Here's the equation:

$$\frac{dN}{dt} = kN - \frac{k}{N_\infty}N^2 \tag{6.1}$$

where N is the population, N_∞ is the maximum size of the population, k is the growth rate and t is the time. Understanding the meaning of the left-hand side of the equation might be tricky if you haven't come across this before. The letters d are not variables at all:

$$\frac{d}{dt}$$

is one expression that symbolizes change or difference, so

$$\frac{dN}{dt}$$

indicates a number that is the change in population with time. We've all met numbers that are rates of change — speed for instance, is the rate of change of distance with time. One way to look at it is to record the population at two close times, and subtract the population that you recorded at the earlier time from what you recorded at the later time. This is done continuously to get the differences at a whole range of times.

The whole equation states that the change in population is equal to the difference between the two terms on the right-hand side of the equation. The first term describes growth and is equal to a parameter known as the growth rate k which is different for each specific case, multiplied by the population. The second term describes the eventual decrease in population and is equal to k divided by N_∞, the maximum size of the population, also known as the *limiting capacity,* and then multiplied by the square of the population.

The nice thing about logistic equations is that there is a mathematical way to solve them (using calculus) so that they are predictive. In other words, logistic equations can be solved analytically to give us a formula for the predicted population at any given time. Unfortunately, k is case-specific, so we cannot *really* predict what's going to happen, especially when using the equation to track the number of people infected by a disease, unless we know this parameter, which we can usually do only in retrospect. Also, we can take steps to quell a pandemic before the whole population gets infected, by isolating the virus or by other means, so the limiting capacity is usually not the total number of people in a country, or city or whatever. Well, there goes our predictive powers! But not all is lost. There is value in doing the math. We can solve the equations and at least understand what's happening at a given moment, i.e. where we are on the curve, or look at the graphs in retrospect to understand what has happened and learn for the future.

The solution for Eq. (6.1) gives us a formula for the total population at any time N_t given its initial value N_0 and its limiting capacity, N_∞. Here it is:

$$N_t = \frac{N_\infty}{1 + \dfrac{N_\infty - N_0}{N_0} e^{-kt}} \tag{6.2}$$

Equation (6.2) looks intimidating but, just like a recipe in a cookbook, all you need to do now is to plug in numbers and calculate the population. Here's how it's done! Let's calculate the rate of infection of COVID-19 in Turkey during the first couple of months of the "first wave", using the logistic equation and compare it to data from the World Health Organization, as it appears on the *Our World in Data* website. First, we'll take N_0, the number of initial COVID-19-positive people as 1000. Why is this not 1? Because in models like these, where we expect a large final population — maybe hundreds of thousands — it might take a long time at the beginning until the infection starts "running". We saw this quite a lot in many countries with COVID-19. So it is usually more accurate to start with a number that is not right at the beginning of the pandemic, yet small enough so as to still count as a "beginning". So, N_0 is taken to be 1000. We'll take the limiting factor, the total number of people that will contract COVID-19, N_∞ by the end of the first wave as 140,000. We know this of course in retrospect. The time t will be counted in days. Following Pelinovsky, we'll choose k to be 0.144 (with units of days^{-1}). Again, this constant was calculated in retrospect.

Now, you can do the calculation yourself! Open a spreadsheet, in Microsoft Excel™, or Apple's Numbers™, or Google Sheets™, or any other spreadsheet program. In one column write the numbers 1 through 60. This represents the days. At the top of the column next to it write the formula for the right side of the equation and drag it down through the rows in the column. The formula in most spreadsheets will look more or less the same. If the column with the days is the first column, it will take up the cells, A1 through A60. The cell B1 will hold the formula:

$$=140000/(1 + ((140000-1000)/1000)*\exp(-0.144*A1)).$$

When dragged down column B, the parameter A1, should change accordingly, to A2, A3, A4 and so on. The last column should be:

$$=140000/(1 + ((140000-1000)/1000)*\exp(-0.144*A60)).$$

Now generate the graph. You'll get the S curve that looks like the black dotted circles in Fig. 6.5. The grey squares in Fig. 6.5 is the real data, taken from the *Our World in Data* website. The similarity is remarkable!

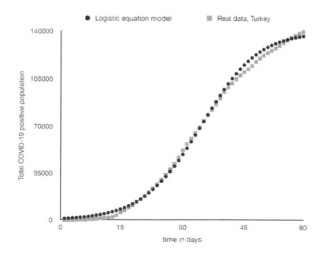

Fig. 6.5   COVID-19 first wave logistic equation prediction vs. real data for Turkey

Unfortunately, this calculation is based on two factors that we can only know in retrospect: the growth factor, k, and the number of people infected at the end of the wave, N_∞. Furthermore, the logistic equation proved quite accurate for some countries, like Turkey, but for other countries it wasn't accurate at all, probably because the notion that the growth rate is constant is wrong, since human behavior "interferes" with pure math! By maintaining social distance, wearing masks, washing hands, lockdowns and the like, we are slowing down the curve, which translates mathematically into changing the value of k with time. Pelinovsky and many others also used Francis Richards's general logistic equation to derive better formulas, and these did work better in some countries. Alas, at least up to the time I'm writing this, no one equation can explain what's happening, even in a single country!

Mathematicians have been instrumental though, by being able to predict what will happen *without* placing restrictions on human behavior — how long will it take for herd immunity to kick in, and what'll be the health price for this (very heavy!). The accumulating data from the different countries, tied to the restrictions placed on the population by the government at different times, help governments learn which limitations are useful in curbing the outbreak and which are not. And there is still the ongoing search for equations that can accurately predict what'll happen considering human behavior. This is an active field in chaos theory and is likely to boom in the coming years.

Math also helps in other aspects of the pandemic. Here is another pandemic-related math puzzle that mathematicians had to deal with:

> *"What is the minimum length of a square room that can hold 5 people each maintaining a distance of 2 meters from each other?"*

One way to solve the puzzle is to place the five people, each at the center of an imaginary circle with a 2-meter diameter, we can rephrase the puzzle mathematically as: what is the minimum length of the side of a square that can pack 5 identical, non-overlapping, 2-meter diameter circles? It's not easy to figure out the minimum configuration — we'll get to that in a moment — so first, I'll show you the answer, the optimal configuration (Fig. 6.6), and then we can calculate the room's size.

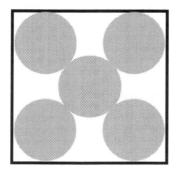

Fig. 6.6   Minimum packing of 5 circles in a square

The diameter of each of the circles is 2 meters. Using algebra, and the Pythagorean theorem (see Chapter 1), we'll use x to mark the length of the side of the square. The length of the diagonal will therefore be:

$$\sqrt{2x^2}. \tag{6.3}$$

The tricky part is working out the length of the two small diagonal segments along the main diagonal between the boundaries of the outer circles and the corners of the square. These segments are equal in size (for symmetry reasons) and can be calculated by drawing a small square encompassing the corner of the square and a quarter of a circle, as shown in Fig. 6.7.

The radius of the circle, which equals the length of the side of the small square, is 1 meter. Using the Pythagorean theorem again, this

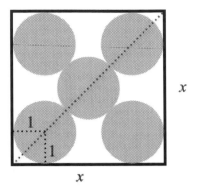

Fig. 6.7   Solving the minimum packing of 5 circles in a square puzzle

time for the small square, we get the length of the segment between the center of the circle and the corner of the square: $\sqrt{2}$ meters. So this minus the radius is the length of the segment we are looking for (the one between the boundary of the circle and the corner of the square): $\sqrt{2}-1$ meters. We can now find x since

$$\sqrt{2x^2} = 6+2(\sqrt{2}-1), \tag{6.4}$$

$$x = \frac{(6+2(\sqrt{2}-1))}{\sqrt{2}}, \tag{6.5}$$

$$x = \frac{(4+2\sqrt{2})}{\sqrt{2}} = 2\sqrt{2}+2, \tag{6.6}$$

and the answer is that the smallest side length of the square room is $2+2\sqrt{2}=4.828...$ meters, giving a total area of approximately 23.314 square meters.

We still have to prove that the given distribution of the circles within the square in Figs. 6.6 and 6.7 is indeed the tightest possible fit. To do this, let's insert a square into Fig. 6.6 whose corners are at the centers of the outer circles, as shown in Fig. 6.7. We'll also divide this square into four quarters, and for simplicity assume that the length of the side of the inner square is one unit (this in effect just changes the scaling of the puzzle but nothing else). Now, imagine that the circles can all move about, and we're trying to find their optimal position. Since there are 4 sub-squares, but 5 points (the centers of the circles), there has to be a sub-square with two points within, so the maximum distance that can possibly be between two points, i.e. the maximum diameter of the circle, is the length of the diagonal of a sub-square, which is, using the Pythagorean theorem, $\frac{1}{\sqrt{2}}$. The shortest distance between two points, on the other hand, is when the circles just touch each other, with a distance of exactly two radii along a straight line connecting their centers, which is also equal to the diameter of the circle, $\frac{1}{\sqrt{2}}$. Since both the minimum and maximum distances are the same, this is also the optimal — so the configuration has to be that in Figs. 6.6, 6.7 and 6.8.

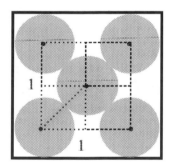

Fig. 6.8   Proof of the optimal packing of 5 circles in a square

The "people in the room" puzzle comes from a well-known topic in math called "packing problems". Packing problems ask questions like:

- How can you optimally pack *n* identical non-overlapping unit circles in the smallest possible square?

- How can you optimally pack *n* identical non-overlapping unit squares in the smallest possible equilateral triangle?

- How can you optimally pack *n* identical non-overlapping unit spheres in the smallest possible regular tetrahedron?

and so on ...

For COVID-19, packing problems allowed some governments to calculate the number of people that can safely fit into a given space. There is no known algorithm, equation or method by which we can find solutions to all packing problems. Indeed, most of the optimal packings nowadays are calculated by computers and cannot even be *verified* by math. Arguably, this field has the greatest number of "open questions", questions that mathematicians have yet to answer. As far as I know, up to date, no one has yet proven the optimal packing of 31 unit circles in a square, or 32 equilateral triangles in a unit circle, or 25 regular hexagons with unit side length in a larger hexagon, and many other packing problems.

Packing problems are a great source of enjoyment for puzzlers. Fitting cubes into rectangular boxes, tiling a given shape with triangular pieces, and many more mechanical puzzles are all packing problems. Given that computer programs cannot solve many of these problems in a decent time, and analytic math cannot do this either, leaves us enthusiasts with the one and only trial-and-error method, which, when coupled with some good old intuition, gives us that great satisfactory "Aha!" moment when we get the last piece in.

The "Aha!" moment — it's when the solution "dawns" on you, a faint smile curves your lips, and you get a real satisfaction from having solved the problem. I love that feeling, and I'm sure you do to. Given that you've read so far, you'll realize that this is really and truly the motive of this book. To give us all some more "Aha!" moments.

Recap

In this final chapter we looked at some of the real-life math puzzles that we had to solve during the 2019–2020 COVID-19 coronavirus pandemic. We started off with a geometric series puzzle, with roots in ancient civilizations. We then took a good look at logistic equations, especially the classic equation of Belgian mathematician, Pierre François Verhulst. We saw how the equation nicely reproduced the data in the first wave of the pandemic in Turkey, and discussed why it failed in some other cases. Finally, we saw how social distancing indoors is akin to packing puzzles, one of the hottest topics in math and computer science.

Here are some of the concepts we saw in this chapter:

- *Base-2 logarithm* — the opposite operation of "taking the power of 2". The base-2 logarithm answers the question: "2 to the power of what" equals the given number. For example, the base-2 logarithm of 8 is 3 because $2^3 = 8$.

- *Generalized logistic function* — a "master equation" from which logistic models can easily be derived for many cases.

- *Geometric series* — a series of numbers where, between every two consecutive numbers, there is a constant ratio.

- *Growth rate* — a case-specific, constant parameter in the logistic equation that describes how quickly the population will grow. The choice of growth rate influences the shape of the sigmoid.

- *Limiting factor* — the expected final population which will always be a constant (at least for some time), i.e. the value of the population when the plateau of the sigmoid is reached.

- *Logistic function* — a differential equation that describes population dynamics, i.e. the rate of change of population with time.

- *Ordinary differential equation* — an equation of the rate of change of one parameter relative to another.

- *Packing problems* — problems that deal with optimally packing shapes within other shapes.

- *Population dynamics* — a mathematical description of how real-life systems behave.

- *Rhind mathematical papyrus* — a historical papyrus discovered by archaeologist Henry Rhind, on display at the British Museum. Written by a scribe named Ahmes in the 16th century BCE, it covers many topics in ancient Egyptian math.

- *Sigmoid* — an S-shaped graph.

- *The Elements* — the 13-book math text written by Euclid and used up until the 19th century BCE.

Challenge Yourself!

1. A wonderful, English nursery rhyme goes like this: "As I was going to St. Ives, I met a man with seven wives. Every wife had seven sacks, every sack had seven cats, every cat had seven kitts. Kitts, cats, sacks, wives, how many were going to St. Ives?" What are two possible answers to this rhyme?

2. The sum of a geometric series of n numbers is given as follows:

$$d + qd + q^2 d + \ldots + q^{n-1} d = \frac{d(1 - q^n)}{1 - q}. \tag{6.7}$$

Find the sum of the following *infinite* geometric series:
(a) 1, 5, 25, 125, …
(b) 1, 0.5, 0.25, 0.125, 0.0625, …
(c) 1, −0.5, 0.25, −0.125, 0.0625, …

3. A meticulous ant starts walking on the table. It walks forwards 32 cm, then turns left and walks forwards 16 cm, then turns left again and walks 8 cm, and so on, indefinitely. After making an infinite number of turns, what is the "as the crow flies" distance from its original starting point? Try and generalize

for any initial distance that it chooses to walk d, and for any factor q. In the problem: $d = 32$ and $q = 0.5$.

Hint: Work out the answer on a grid with Cartesian coordinates.

4. Estimate the world population in 2100, if in 2020 there were 7.8 billion people living, the growth factor, k, is 0.011 (years^{-1}), and the limiting capacity of Earth is 20 billion people.

5. A country with a population of 20,000,000 people reaches herd immunity when about half the population is infected with a certain virus. On January 1st, one person infected with the virus arrives in the previously uninfected country and the virus begins to spread. How many people will be infected 30 days after his arrival, if the known growth rate factor is 0.4 (days^{-1}), and we assume that the limiting capacity is equal to the number of people infected when herd immunity is achieved?

6. Figure 6.9 shows the optimal way to pack two unit-circles in an equilateral triangle. What is the length of the side of the triangle?

Fig. 6.9   Optimal packing of two unit-circles in an equilateral triangle

7. What is the optimal way to pack three unit-circles in an equilateral triangle and what is the length of the side of the triangle?

Solutions

1. The trivial answer is 1, assuming that the entourage was *returning* from St. Ives. If, however, we met up and walked together to St. Ives, apart from myself there are another $7^0 + 7^1 + 7^2 + 7^3 + 7^4$, a geometric series, with a sum of 2801. Notice the resemblance to Ahmes's geometric series riddle!

2. (a) The sum is infinity.
   (b) Notice that when the geometric factor q is less than 1, when n approaches infinity, $q^n = q^0 = 1$, and Eq. (6.7) becomes

$$d + qd + q^2 d + \ldots = \frac{d}{1-q} \qquad (6.8)$$

   and the answer is $\frac{1}{0.5} = 2$.
   (c) Using Eq. (6.8), the answer is: $\frac{1}{1.5} = \frac{2}{3}$.

3. First, we need to calculate where the ant will be along the x axis (horizontally), and where it will be along the y axis (horizontally). This breaks down the problem into two geometric series:

   The x axis:

$$d - q^2 d + q^4 d - q^6 d - \ldots = \frac{dq}{1+q^2}.$$

   The y axis:

$$qd - q^3 d + q^5 d - q^7 d - \ldots = \frac{dq}{1+q^2}.$$

   The "as the crow flies" distance is, according to the Pythagorean theorem,

$$\sqrt{a^2 + b^2} = \sqrt{\frac{d^2(1+q^2)}{(1+q^2)^2}} = \frac{d}{\sqrt{1+q^2}},$$

which, when using $d = 32$, $q = -0.5$, equals

$$\frac{32}{\sqrt{1+\left(-\dfrac{1}{2}\right)^2}} = \frac{64}{\sqrt{5}} = 28.622 \text{ cm.}$$

4. Using Eq. (6.2), and taking $N_0 = 7.8$ (billion), $N_\infty = 20$ (billion), $t = 80$ (years) and $k = 0.011$ (years^{-1}), we get:

$$N_{80} = \frac{20}{1+\dfrac{20-7.8}{7.8}\times e^{-0.011\times 80}} = 12.13 \text{ billion people.}$$

5. Using Eq. (6.2), and taking $N_0 = 1$, $N_\infty = 10{,}000{,}000$, $t = 30$ days and $k = 0.4$ (days^{-1}), we get:

$$N_{30} = \frac{10000000}{1+9999999\times e^{-0.4\times 30}} = 160{,}148 \text{ people.}$$

6. Figure 6.10 shows the answer. The base of the triangle is equal to two radii plus the two segments between the points below the centers of the circles and the vertices of the triangle (x), or $2 + 2x$. x is the base of a small 30–60–90 triangle, whose

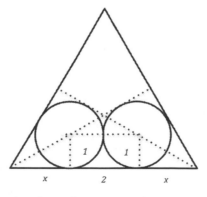

Fig. 6.10   Optimal packing of two unit-circles in an equilateral triangle

hypotenuse extends from a vertex of the triangle to the center of the nearest circle, and whose altitude is 1 unit (the radii of the circle). Note that the angle bisectors intersect the centers of the circles, since each circle is inscribed in a triangle whose area is half the area of the equilateral triangle. In a 30–60–90 triangle with unit altitude, the long base x is equal to $\sqrt{3}$, so the length of the side of the packing triangle is $2+2\sqrt{3}$.

7. From symmetry reasoning alone, it is easy to see that one more circle fits in the triangle in Fig. 6.10, so the solution is the same as the previous question, and the length of the side of the triangle is $2+2\sqrt{3}$.

Bibliography and Further Reading

Ahmes (n.d.). The Rhind mathematical papyrus. *The British Museum.* https://www.britishmuseum.org/collection/object/Y_EA10058

Chace, A. B. (1927). *The Rhind Mathematical Papyrus.* The Mathematical Association of America, Oberlin, OH.

Friedman, E. (n.d.). Packing equal copies, *Erich's Packing Center.* https://erich-friedman.github.io/packing/index.html

Nurmela, K. J., & Östergård, P. R. J. (1997). Packing up to 50 equal circles in a square, *Discrete & Computational Geometry* **18**(1), pp. 111–120.

O'Connor, J. J., & Robertson E. F. (1999). Euclid of Alexandria, *MacTutor History of Mathematics Archive.* https://mathshistory.st-andrews.ac.uk/Biographies/Euclid/

O'Connor, J. J., & Robertson E. F. (2014). Pierre François Verhulst, *MacTutor History of Mathematics Archive.* https://mathshistory.st-andrews.ac.uk/Biographies/Verhulst/

Pelinovsky, E., Kurkin, A., Kurkina, O., et. al. (2020). Logistic equation and COVID-19, *Chaos, Solitons & Fractals*, **140**, 110241.

Richie, H., Ortiz-Ospina, E., Beltekian, D., et. al. (2020), Coronavirus Pandemic (COVID-19), *Our World in Data*. Retrieved December 26, 2020, from https://ourworldindata.org/coronavirus

Figure Credits

Fig. 1.1   Lewis Carroll, 1863 photograph by Oscar G. Rejlander
[Credit: Photographer Oscar G. Rejlander,
https://commons.wikimedia.org/wiki/File:Lewis Carroll 1863.jpg,
public domain]

Fig. 2.11   Iteration of Wolfram's "Rule 30" for a large number of time steps
[Credit: Zhiming Wang,
https://github.com/zmwangx/rule30/blob/master/images/rule30.png]

Fig. 3.8   Portrait of Leonhard Euler by Jakob Emanuel Handmann (1753)
[Credit: Artist Jakob Emanuel Handmann,
https://commons.wikimedia.org/wiki/File:Leonhard_Euler.jpg, public domain]

Fig. 3.9   Albrecht Dürer's engraving, Melancholia I [Credit: Artist Albrecht Dürer,
https://commons.wikimedia.org/w/index.php?curid=29675382,
public domain]

Fig. 3.10   The magic square in Melancholia I [Credit: Artist Albrecht Dürer,
https://commons.wikimedia.org/w/index.php?curid=752363, public domain]

Fig. 3.14   Sudoku the Giant [Credit: Timothy57,
https://commons.wikimedia.org/w/index.php?curid=34167805,
public domain]

Fig. 3.15   Solution to Sudoku the Giant [Credit: Timothy57,
https://commons.wikimedia.org/w/index.php?curid=52717217,
public domain]

Fig. 5.21   First four iterations of the Menger sponge
[Credit: Solkoll, https://commons.wikimedia.org/w/index.php?curid=64192,
public domain]

Fig. 5.23   A simple fractal tree [Credit: Saisundar.s,
https://commons.wikimedia.org/wiki/File:Simple Fractals.png,
https://creativecommons.org/licenses/by-sa/4.0/legalcode]

Fig. 6.1   Sissa ibn Dahir (artist's impression) [Credit: Thiago Cruz,
https://commons.wikimedia.org/wiki/File:Lahur_Sessa_by_Thiago_Cruz.jpg,
public domain]

Fig. 6.2   A fragment of the Rhind Papyrus [Credit: Paul James Cowie,
https://commons.wikimedia.org/w/index.php?curid=6943889, public domain]

Fig. 6.3   Pierre François Verhulst [Source:
https://commons.wikimedia.org/wiki/File:Pierre_Francois_Verhulst.jpg,
public domain]

Printed in the United States
by Baker & Taylor Publisher Services